From an Ambrotype by E. F. Moore, Wellsburg.

THE LIFE AND TIMES
OF
PATRICK GASS,

NOW SOLE SURVIVOR

OF THE OVERLAND EXPEDITION TO THE PACIFIC,
UNDER LEWIS AND CLARK IN 1804-5-6;

ALSO,

A SOLDIER IN THE WAR WITH GREAT BRITAIN, FROM
1812 TO 1815, AND A PARTICIPANT IN THE
BATTLE OF LUNDY'S LANE.

TOGETHER WITH

GASS' JOURNAL OF THE EXPEDITION CONDENSED;

---AND---

SKETCHES OF SOME EVENTS OCCURRING DURING THE
LAST CENTURY IN THE UPPER OHIO COUNTRY,
BIOGRAPHIES, REMINISCENCES, ETC.

BY J. G. JACOB.

Lone Wolf Press
Mansfield Centre, Connecticut
2000

Lone Wolf Press, P.O. Box 373, Mansfield Centre CT 06250
U.S.A.

ISBN 1-57 898-274-X

© Lone Wolf Press

All rights reserved. No new contribution to
this publication may be reproduced, stored
in a retrieval system, or transmitted, in any form or by
any means, electronic, mechanical, photocopying,
recording, or otherwise, without the prior permission of
Lone Wolf Press

This work is a facsimile edition of
The Life and Times of Patrick Gass
by J.G. Jacob
Published in Wellsburg, VA in 1859

Printed in the United States of America
On 100% Acid-Free Paper

THE
LIFE AND TIMES
OF
PATRICK GASS,

NOW SOLE SURVIVOR

OF THE OVERLAND EXPEDITION TO THE PACIFIC,
UNDER LEWIS AND CLARK, IN 1804-5-6;

ALSO,

A SOLDIER IN THE WAR WITH GREAT BRITAIN, FROM
1812 TO 1815, AND A PARTICIPANT IN THE
BATTLE OF LUNDY'S LANE.

TOGETHER WITH

GASS' JOURNAL OF THE EXPEDITION CONDENSED;

—AND—

SKETCHES OF SOME EVENTS OCCURRING DURING THE
LAST CENTURY IN THE UPPER OHIO COUNTRY,
BIOGRAPHIES, REMINISCENCES, ETC.

BY J. G. JACOB.

JACOB & SMITH,
PUBLISHERS AND PRINTERS, WELLSBURG, VA.
1859.

Entered according to Act of Congress, In the year 1856, by
J. G. JACOB & J A. SMITH,
In the Clerk's Office of the District Court of the United States,
for the Western District of Virginia.

PREFACE.

The design of preparing the following pages, was conceived during a period of leisure, and originally contemplated only a series of sketches for a weekly newspaper; but as the subject grew under the hand of the author, the original design was abandoned and the work assumed its present form. For the sin of adding another to the innumerable books, we have no other apology to offer. A curiosity was expressed to have on record the adventures of its hero, and his eventful career promised material for an interesting story.

We have done the best possible with our materials. If there be merit in the work, the reader will not be slow to discern it; if there be none, the public will not be backward about demonstrating that fact.

The biography of its citizens is the history of a nation; and we trust that the reputation of history will not suffer from one departure in permitting the humble biography of a hero of the ranks, to stand by the side of those of the great and titled, who have been by common consent, but very untruthfully, held up as the exponents and exemplars of the nation.

Patrick Gass, is the type of the self reliant, adventurous American citizen and soldier, who has carried the eagles of the Republic in triumph through three wars; and planted the olive branch on the highest pinnacle of the temple of Peace.

The concluding pages of our work will we hope, prove acceptable to a numerous class to whom the material there used is no novelty, but who may perhaps be gratified in having it systematically and conveniently arranged. It, of course does not pretend to be a full history of the events of the last eighty years; but as far as it goes, will we think, be found reliable; and may serve to assist some future historian. We have to regret that our allotted space is too small to allow the insertion of more reminiscences, or to permit as much detail as we could desire, on the subjects we have attempted. Several sketches, and other material, the result of considerable research, have been thus omitted, while others have been so abreviated as to be unsatisfactory. We had designed giving a detail of the Whiskey Insurrection, also a history of the settlement of Wellsburg and Wheeling, an account of the Railroad Era, and various other matters for which we had the material provided, but inexorable necessity forbade it.

Although great care has been observed, it is not improbable also that some inaccuracies may be found. Authorities themselves, although contemporary, often differ; hence, it is the more difficult for subsequent writers to be exact. Should this little volume be received with favor, a subsequent publication may correct its errors and amplify its details.

To those who have kindly contributed matter or otherwise assisted us, we return our sincere acknowledgements.

With these prefatory remarks, the book is submitted to a generous public.

THE AUTHOR.

Wellsburg, January, 1859.

CONTENTS.

Boyhood and Youth,	PAGE 11
Moving to the West,	14
Wellsburg in 1790,	19
Gass' Services on Yellow Creek,	21
Recollections of Lewis Wetzel,	22
Flatboatmen,	24
Thomas and Solomon Eury,	26
James Buclvinan, Senior and Junior,	29
Enlistment in the French War,	31

OVERLAND JOURNEY TO THE PACIFIC--CHAP. II

Death of Sergeant Floyd,	42
Encampment at Fort Mandan,	57
Sickness of the Party,	86
Flath, ad Indians,	88
The Pacific in sight,	93
Departure for home,	100
Arrival of the party at St. Louis,	106
Travel through the States,	108
Lewis and Clark-subsequent history,	110
Pacific Railroad,	114
Mormons-Fremonts' and Gunnison's disasters,	115
Beckwith's Pass,	118
Distances and Route traveled by Lewis & Clarke,	119
Western Indians-their present condition,	120
Troubles in Kansas,	135
Publication of Gass's Journal,	140

THE WAR OF 1812.--CHAP. III.

Right of Search and Impressment,	146
Gass enlists for the war,	147
Trip from Kaskaskia to Pittsburg,	149
Niagara Campaign,	151

CONTENTS.

Battle of Chippewa,	151
Battle of Queenstown,	156
Battle of Lundy's Lane,	159
Investment of Fort. Erie,	162
Bombardment and attack of Fort Erie,	163
Sortie from Fort Eric,	167
Reminiscences of Campaign	170
Discharge from service	175
Courtship and Marriage,	177
Compensation from Government,	179
Pension Laws,	180
Old Soldiers Convention,	184
President Pierce's Address,	187
Resolutions of the Convention,	188
Conclusion of Biography,	193

CIVIL HISTORY--PART II.

The Upper Ohio--French and English Pretensions,	195
Washington's First Expedition,	195
First English Settlement,	197
Washington's Second Expedition,	198
Washington's Capitulation,	200
The Ohio Indians,	203
Braddock's Expedition--Defeat and Death,	211
Pontiac's War--Col. Boquet,	222
William Pitt,	224
Forbes' Expedition against Fort Duquesne,	225
Grant's Defeat--Fall of Fort Duquesne,	225
Early Boundary Disputes,	227
Pennsylvania and, Virginia State line,	228
Patents--Location--Litigation,	229
Lord Dunmore--Col. Connelly,	230
Cessation of the Dispute,	231
Final Settlement,	232
The Panhandle,	234
West Liberty as a Seat of Justice,	236
Early Settler's Names--Enterprise,	240
Weighty Characters,	244
Early Religious Inclinations--Presbyterianism,	245

CONTENTS.

Origin of Canonsburg and Washington Colleges,	246
Redstone Presbytery,	247
Origin of Camp Meetings--Methodism,	248
Lorenzo Dow,	250
Rev. James Finley,	251
Stone Meeting House on Short Creek,	252
Baptist Church,	253
Episcopal Church--Dr. Doddridge,	254
Schools and Colleges,	258
Alexander Campbell,	261
Bethany College,	263
West Liberty Academy.--Wellsburg Seminary,	266
Free Schools,	267
National Road,	268
Philip Doddridge,	273
Wellsburg and Washington Turnpike,	275
The Whiskey Insurrection,	277
Adam Poe and Big Foot,	279

LIFE AND TIMES

OF

PATRICK GASS.

PATRICK GASS, the subject of our memoir, is a hero in humble life. He cannot trace his descent down a long line of illustrious ancestors nor has his unpretending name been ennobled by courtly favor, or, by brilliant deeds in diplomacy or war; yet was he born in stirring times. His cradle was literally rocked amid the storms of the revolution and the days of his manhood extend through a most eventful era. In the events of his day he has performed although an humble, yet a not unimportant part, and perhaps, as well deserves a niche in the temple of fame as many a one to whom fortune has been more propitious. But it was his fate to serve, rather than to command; and as the ways of the world will have it, so we must regard him, in his subordinate capacity, much as we might wish that his position was, for our purpose, a more conspicuous one.

He first saw the light, June 12th, 1771, at Falling Springs, Cumberland county, near what is now Chambersburg, in Franklin county, Pennsylvania. At this date, October, 1858, he is still living, a hale, hearty old man, with the apparent promise of many years of life yet to come. His freedom from the infirmities of an age so advanced is not the least remarkable characteristic of his history. It cannot be attributed to extraordinary freedom from exposure nor to excessive sobriety or regularity of habits; on the contrary, the reverse appears to be the fact. In his early manhood, he exposed himself during twenty years of military service, to all the casualities of the soldier, both in actual service and in camp, to disease, privation and danger in every form: and after his retirement from service he was as addicted to a weakness for strong drink, which, he for many years, indulged to an extent that few men could have born with impunity. Yet, through all, he led apparently, a charmed life and still lives a monument of God's mercy and of the enduring power of the human constitution. Although his years considerably outnumber those usually allotted to man, he preserves all his faculties in a remarkable degree. Physiologically considered, he is of the build most favorable for longevity. In stature, somewhat low, never having in his best estate, exceeded five feet seven, stoutly and compactly built, broadchested and heavy limbed, yet lean, sprightly and quick of motion, his physique exhibits the amplest and most compact developement of vital force of which the human frame is capable. Although now, somewhat bowed and slightly crippled with the rheumatism, he is a remarkably alert and active walker and can make the four miles from his residence to Wellsburg, in about as good time as most of those of one fourth his years. Indeed, he prides himself upon his pedestrianism and delights to, jibe the "pups", as he is pleased to call the youth of these degenerate times upon their effeminacy in this respect. His intellectual development is good. His eye is bright and lively, his mind active and discriminating, his memory of events of fifty years ago, accurate and reliable, and the general expression of his countenance intelligent and reflective. He is naturally a man of good sound sense, very observant, and disposed to turn his observations to practical account.

    He seems to regard the retention of his faculties with a warrantable pride, and we have no doubt still considers himself competent for a fair day's march. We have heard him declare, with all the enthusiasm of a conscript, his readiness to shoulder arms again in the service of his country, should occasion require it, and have no doubt whatever of his sincerity. His parentage was Irish, which probably accounts for his patriotic proclivities as he seems to inherit the hatred of British domination so common to the sons of the Green Isle, and which appears to be transmitted by hereditary descent.

    Of his boyhood not much is to be said more than might be said of the boyhood of other men. After several removes on the part of his father, a notable one was over the South Mountain into Maryland, in 1775, just at the time that the revolutionary contest was assuming the stage of a civil war. There is no doubt but that the lessons of abhorrence to British tyranny, early and insensibly impressed upon his mind at this time, adhered to him through life and exerted an influence on his after career. From 1777 to 80, he resided with a grandfather ostensibly for the purpose of attending school, but the facilities of that day, must have been extremely poor for he informs us that the total effective amount of his school education extending to reading, writing and cyphering, was acquired in 19 days, and that, after he had come to the years of manhood. We have heard him regret that his early education had not been better, for he seems in his old days to entertain the idea that he might, with his opportunity and mental and bodily energy have attained an eminence among the great men of the nation. However, his case was no anomaly in his day, the means of acquiring an education being extremely limited and very few of his cotemporaries being further advanced than the commonest rudiments of English learning. He, however, like them took his lessons from men and things and made up for lack of book learning, by close observation and shrewd reasoning.

His next recorded move was in 1780, rendered memorable by the hard winter of that year, being the most severe almost in the history of this country. At this time the prospects of the American cause seemed almost hopeless, and it may well be called the dark day of the revolution. The worn army of Washington was hutted among the short hills of Morristown, famishing some times for want of food, often naked and cold, the continental currency had depreciated until $40, would scarcely buy a bushel of corn; the commissariat was sadly deranged, a general gloom of despair settled over the hopes of the Patriots, and as if Providence, too, had combined with their enemies, the winter of this year set in with a severity that was entirely unprecedented. The harbor of New York was frozen up and the British fleet stationed there to protect it from the Americans, was rendered useless, so that Kniphausen debarked the sailors and marines, and organizing them into land forces, prepared them to defend the city against a contemplated attack by Washington, over the frozen waters of the bay. The Schuylkill at Philadelphia, was frozen so as to bear the heaviest artillery and wagons. The suffering of the American soldiery was intense. And not only they suffered for want of the common necessaries of life, but the population of the Jerseys and of Eastern Pennsylvania, the debatable ground between the British and Americans, harrassed and foraged over and over again by both parties were also impoverished and reduced to want. This state of affairs was not confined to the 'debatable ground' but extended throughout the whole seaboard, where was the theatre of war. It induced emigration toward the back country whence, while the family was secure from disturbance, the yeomen could sally forth to the defence of the country. Considerations of this kind influenced the elder Gass, with many others, to move toward the west, and no doubt the contrast between the pitiable condition of the patriot army and the well fed and well housed hirelings of the enemy had its effect upon the plastic mind of the boy of nine years of age as well as upon others, to be developed in after years. There is no doubt but that one effect of the harassing nature of the war of the revolution, was to diffuse population and thus hasten the settlement of the country, and thus under the blessing of Providence proved the cause of its remarkable development.

Accordingly in 1782, his father moved the family to the "west" then located on the further side of the Allegheny mountains, but since removed, year by year, until the name "west" has now become a phrase of very indefinite meaning. They encountered great hardships on the road, of which the following graphic reminiscence taken from the pages of "Old Redstone," will convey an idea. "My father's family," says the author, 'was one of twenty that emigrated from Carlisle, and the neighboring country, to Western Pennsylvania, in the spring of 1784. Our arrangements for the journey, would, with little variation be descriptive of those of the whole caravan. Our family consisted of my father, mother, and three children, (the eldest one five, the youngest less than one year old,) and a bound boy of fourteen. The road to be travelled in crossing the mountains was scarcely, if at all, practicable for wagons. Pack-horses were the only means of transportation then, and for years after. We were provided with three horses, one of which my mother rode, carrying her infant, with all the table furniture and cooking utensils. On another were packed the stores of provisions, the plough irons, and other agricultural tools. The third horse was rigged out with a pack-saddle, and two large creels made of hickory withes, in the fashion of a crate, one over each side, in which were stowed the beds and bedding, and the wearing apparel of the family. In the centre of these creels there was an aperture prepared for myself and sister; and the top was well secured by lacing, to keep us in our places, so that only our heads appeared above. Each family was supplied with one or more cows, which was an indispensable provision for the journey. Their milk furnished the morning and evening meal for the children, and the surplus was carried in canteens for use during the day.

   Thus equipped, the company set out on their journey. Many of the men being unacquainted with the management of horses, or the business of packing, little progress was made, the first day or two. When the caravan reached the mountains, the road was found to be hardly passable for loaded horses. In many places, the path lay along the edge of a precipice, where, if the horse had stumbled or lost his balance, he would have been precipitated several hundred feet below. The path was crossed by many streams, raised by the melting snow and spring rains, and running with rapid current in deep ravines. Most of these had to be forded, as there were no bridges and but few ferries. For many successive days, hair-breadth escapes were continually occurring; sometimes, horses falling; at other times, carried away by the current, and the women and children with difficulty saved from drowning. Sometimes, in ascending steep acclivities the lashing of the creels would give way, and both children and creels tumble to the ground, and roll down the steep, until arrested by some traveller of the company. In crossing streams or passing places of more than ordinary difficulty in the road, mothers were often separated from some of their children for many hours.

The journey was made in April, when the nights were cold. The men who had been inured to the hardships of war, could with cheerfulness endure the fatigues of the journey. It was the mothers who suffered; they could not, after the toils of the day, enjoy the rest they so much needed at night; the wants of their suffering children must be attended to. After preparing their meal they lay down, with scanty covering, in a miserable cabin, or, as it sometimes happened, in the open air, and often, unrefreshed, were obliged to rise early, to encounter the fatigues and dangers of another day.

As the company approached the Monongahela, they began to separate. Some settled down near to friend and acquaintances who had preceded them. About half of the company crossed the Monongahela, and settled on Chartier's creek, a few miles south of Pittsburgh, in hilly country, well watered and well timbered. Settlers' nights to land were obtained on very easy terms. My father exchanged one of his horses for a tract, (bounded by certain brooks and marked trees,) which was found on being surveyed several years after, to contain 200 acres.

The new-comers aided each other in building cabins, which were made of round logs, with a slight covering of clapboards. The building of chimneys and laying of floors were postponed to a future day. As soon as the families were all under shelter, the timber was girdled and the necessary clearing made for planting coat, potatoes, and a small patch of flax. Some of the party were despatched for seed. Corn was obtained at Pittsburgh; but potatoes could not be procured short of Ligonier valley, distant three days' journey. The season was favorable for clearing; and, by unremitting labor, often continued through a part of the night, the women laboring with their husbands in burning brush and logs, their planting was seasonably secured. But, while families and neighbors were cheering each other on with the prospect of an abundant crop, one of the settlements was attacked by the Indians, and all of them were thrown into the greatest alarm. This was a calamity which had not been anticipated. It had beet confidently believed that peace with Great Britain would secure peace with her Indian allies. The very name of Indian chilled the blood of the late emigrants; but there was no retreat. If they desired to recross the mountains, they had not the provisions or means, and had nothing but suffering to expect, should they regain their former homes. They resolved to stay.

The frontier settlements were kept in continual alarm. Murders were frequent, and many were taken prisoners. These were more generally children, who were taken to Detroit, (which, in violation of the treaty, continued to be occupied by the British,) where they were sold. The attacks of the Indians were not confined to the extreme frontier. They often penetrated the settlements several miles, especially when the stealing of horses was a part of their object. Their depredation effected, they retreated precipitately across the Ohio. The settlers for many miles from the Ohio, during six months of the year, lived in daily fear of the Indians. Block houses were provided in several neighborhoods for the protection of the women and children, while the men carried on their farming operations, some standing guard while the others labored. The frequent calls on the settlers to pursue marauding parties, or perform tours of militia duty, greatly interupted their attention to their crops and families, and increased the anxieties and sufferings of the women. The general government could grant no relief. They had neither money or credit. Indeed; there was little but the name in the old confederation. The State of Pennsylvania was unable to keep up a military force for the defence of her frontier. She had generously exhausted her resources in the struggle for national independence. Her Legislature however, passed an act granting a bounty of one hundred dollars on Indian scalps. But an incident occurred which led to the repeal of this law before trio termination of the war.

The Gass family, however, reached the forks of Yough without extraordinary incident, in 1784, and immediately proceeded to locate near Uniontown, then called Beasontown. Their stay however was but short at

Beasontown, for in the ensuing year they again pulled up stakes and removed their household altar to Catfish Camp where Washington now stands. Catfish at that day was a bold stream of many times its present dimensions and indeed, the diminution of the streams is one of the most singular incidents conected with the settlement of this country. The stream in question, has dwindled from a powerful stream to an insignificant brook and we have before us an interesting instance, in point, pertaining to its near neighbor, Chartiers on the authority of Mrs. Jane C. Patterson, relict of Robert Patterson, who died near Pittsburgh in 1858, in her 80th year. Her biography as published in the Pittsburgh Advocate and Journal, states: "She well remembered the building by her father, of the old mill in Canonsburg, one of the first in all the west, and to which the farmers from a great distance around brought their grain. On one occasion, advantage was taken of a high stage of water to send a boat, freighted with barrels of flour almost from the floor of the mill by the tortuous course of the Chartiers, at that time unobstructed by other dams, to the Ohio, and so onward to New Orleans. The clearing up of the country for more than half a century, may possibly account for the present thread-like appearance of the stream, which certainly does not suggest the idea that Canonsburg was once the head of navigation."

Catfish took its title from being the head quarters of a noted Indian chief of that name. His cabin was located on the run about where the steam grist mill now stands. Catfish Camp, was also prominent in early, times from being a sort of half way house between the Monongahela and the Ohio. A regular path existed in those days from Redstone by the waters of Buffalo and Wheeling creeks, to the Ohio at Wellsburg and Wheeling, much travelled by the emigrants as well as by the Indians, and as Catfish was about 24 miles from either river, it was a convenient stopping place, and became, generally known to the settlers and scouts as a rendezvous. It must be borne in mind that travelling in those days was very different from travelling now. The journey from eastern Pennsylvania to Redstone Old Fort, in 1785, was fully equal in magnitude to a trip now-a-days, to Oregon and back. The intermediate country, between Redstone and the Ohio was not only rugged and broken, but was peculiarly infested with Indians. The roads, where roads there were any, almost uniformly followed the highest ridges, so as to avoid any danger of a surprise that might occur by keeping along the ravines. This made the traveling safer but much more difficult. The adventurer, who had safely crossed the Laurel Hills, passed through the "shades of death" and seen the big pines and deep snows of the mountains and finally dared the Indian infested woods at the Ohio, was on his return a hero of no ordinary renown. From Catfish camp, Patrick directed his explorations into the surrounding country: and he gives us his impressions of Wellsburg as the site appeared to him in 1790. The ground was swampy in parts and covered with a dense growth of sycamore, walnut, sugar and wild plum trees. There was at that day but one building to be seen, that was a log house on the lower end of the bottom near midway then, between the river and the hills. It was built and many years occupied by Alexander Wells, and in 1858, is still standing and is the property we believe of Wm. Daten. It has been weatherboarded and a large stone chimney since added to it on the outside. What is now the bar, at the mouth of the creek, was then a high bottom, thickly and luxuriantly covered with a heavy forest growth, and the bottom itself, north of the creek, was much wider than present; hundreds of acres having gone into the river since the occupation of the country by the whites. Indeed at an early day, serious apprehensions were entertained that the entire town site would gradually wear away;and about the year 1830, money was raised by Lottery authorised by the state, to the amount of some $25,000, to build walls to protect the river banks. The privilege was sold to a lottery company, and the proceeds appropiated to the construction of the heavy walls which at present extend along the front of the town. Mr. John Gilchrist, now of Wheeling, was one of the contractors, perhaps the principal one, and the work was completed, or the money expended about the year 1836. The walls have only partially answered their purpose, though they have saved the banks to a considerable extent. At the extreme point of the bar is a stone about ten feet long, of irregular width, known as the "Indian Rock," which in early times bore a number of Indian hieroglyphics and upon which tradition states, George Washington inscribed his name in one of his early journeys to the west. The marks whatever they were, have long since been worn out or cut out by ambitious individuals who have covered its surface with their own initials to the displacement of the "Indian signs." The appearance of the rock marks low water in the river.

At Catfish Camp, Patrick remained on the farm leased by his father for a considerable period, during which time he made several trips over the mountains to Mercersburg and Hagerstown, for salt, iron, & c., which in those days had to be packed on horses--200 pounds of iron or two bushels of salt being the usual burden of a horse. Considering the almost absolute absence of roads, these excursions were attended with considerable labor and not a little peril; but they suited his roving and venturesome spirit admirably, and helped to develope a disposition for travel, that finally culminated in the then unheard of enterprise of an overland journey to Oregon of which we shall have more to say hereafter.

The year 1790, was remarkable for a drouth in the Catfish country, and Patrick came to Charlestown for corn which as he informs us he procured from Mr. Macfarland, the Surveyor who laid out the town, and who subsequently committed suicide by shooting himself.— His corn he took to "Moore's Mill," on Buffalo, got a due bill for the same and returned to Catfish, thus travelling 50 miles or more to mill and back.

What would the farmers of this day with their steammills, turnpikes, railroads and steamboats think if compelled to such a resort for their daily bread, yet such inconveniences were the rule rather than the exception in those days.

About this time having attained his majority, our hero began to feel a longing for the glories of war, and the next we hear of him is in 1792, when he was stationed under Capt. Caton, at Yellow Creek, to help guard the frontiers against the incursions of the Indians, who had been for a long time troublesome, and who were at this time particularly elated with their success in defeating Gen. St. Clair and his army in the November preceding. At this time there was felt the greatest apprehension on account of the Indians, to chastise whom, and effectually conquer a peace, Gen. Anthony Wayne was sent out with a considerable force by the Government, and the militia all along the frontier was drafted into actual service. Patrick on this occasion was serving in lieu of his father who had been drafted. He was himself drafted shortly after leaving Yellow creek, and stationed at Bennett's Fort, three miles from Wheeling, on Wheeling Creek. He does not appear to have been engaged in any actual fighting with the Indians, having been occupied with garrison duty, but deserves not the less credit on that account, for to a man of his temperament the confinement of a fort was more irksome than the hardships of an active campaign.

Shortly after this, in August 1792, the Indians received a decisive check at the hands of Gen. Wayne, in their total defeat on the Miami, which virtually and forever put a close to Indian difficulties in this region. Emigration to the west which had almost entirely ceased during the previous few years commenced again with renewed activity after the peace of Grenville, in 1796, and under the auspices of the Ohio company and the leadership of the veteran officers of the revolutionary war, the Ohio country filled up with great rapidity, and with a class of settlers preeminently qualified for laying broad and deep the foundations of a great and prosperous state.

While stationed at Wheeling Creek, Mr. Gass had an opportunity of seeing the noted Lewis Wetzel and also Capt. Samuel Brady, names common as household words, to all familiar with our early annals. They were then acting as scouts, in which capacity they rendered the infant settlements most effective service. Wetzel is described as a tall, black visaged, slenderly built man, with remarkably keen eyes; and history gives him credit for a deadly and inveterate hatred of the red-skins, of whom he sent more to their last account, than perhaps my other one man of his or any other day.

A noted instance of his success in circumventing the "red skins," is given in the romantic story of " Old Cross Fire," which has more than once gone the rounds of the papers and is substantially true; and to this day the spot is pointed out where an Indian, having concealed himself among the rocks on the precipitous bank of the creek near Wheeling, and decoyed and shot several white men, by imitating the noise of a turkey, was himself killed by Wetzel, while in the very act of "gobbling" for a victim. The identical rock behind which the Indian was concealed was in existence about three-fourths of a mile from the month of Wheeling creek, until a few years ago, when it was split up for building purposes, and sold by the owner of the ground. It is to be regretted that the rock was not preserved, as a standing memorial of the Indian times, and a monument to the brave and intrepid hunter, who has given his name, however, to a county in our good old commonwealth. Had it been named Wetzel's rock, it would have remained an object of interest for ages perhaps; and thus effectually perpetuated his name, and proven a source of gratification to those who would desire justice done to the memory of the pioneers. Like many other men of his contemporaries, Wetzel had private injuries to revenge upon the Indians, and his hatred of them was bitter and relentless. They murdered several of his kindred, and he himself with an elder brother had been prisoners among them when boys, and effected their escape by extraordinary address and courage, and he vowed eternal enmity. Like too many men of his class he had somewhat loose ideas of the sanctity of laws and treaties with Indians, when they interfered with the gratification of his vengeance; and it is reported of him, that he scrupled not to take a shot when occasion offered, even though in time of peace. Of course, such conduct was not only discreditable to the good faith of the whites, but was vitally dangerous to their security from Indian vengeance. He was warned and admonished of the danger to the peace of the settlements of such an uncompromising hostility, and was finally arrested in Ohio, and imprisoned on charge of murder, for shooting an Indian in time of peace. He would doubtless have been visited with the rigors of the law, but popular sympathy was in his favor. The whole country side flocked to the place of his confinement demanding his discharge, under penalty of demolishing the jail and delivering him by force, and the result of the demonstration was, that he was by some unaccountable means released.—After this adventure, tradition speaks of no more filibustering on his part and his subsequent career is involved in doubt, though the most probable story states that he engaged in flatboating on the river, became dissipated and died miserably in New Orleans sometime about the beginning of the present century.

The flatboatmen of that day were an extraordinary race, aptly denominated in the Mike Fink dialect as half horse and half alligator. They were a reckless, frolicking set, not generally burdened with any over-supply of conscience, and yet endowed with a rough sense of honor among themselves and toward their employers. However piratically disposed toward the squatters along the banks of the rivers, and toward outsiders generally; and however ready to engage in broils and to risk their lives for trivial insults, or even for bravado, in the sanguinary fights of their day, they could yet be safely entrusted with uncounted sums of money, and would fight to the death in defence of their comrades or their employers' property. They were composed generally of the restless borderers, who, as in all new countries, prefer a life of excitement and hardship, so that it be coupled with freedom, to one of settled comfort and constraint. Wetzel was a man of this temperament, and it is highly probable that he became an adept in the rough features of boating, and as history is silent, we may reasonably conclude that his career was like that of most of his comrades. The life they led was a precarious one; leaving out of the question the dangers of their calling from accident and disease, the exposure and habitual dissipation so common among them, very generally cut short their careers. Nevertheless, there was an excitement about it which was very attractive to the youth of that day.

The produce of this section was at that day transported by flatboats, batteaux and similar floating craft, to the New Orleans market, then the only outlet for the surplus production, and as the risk was great and the labor severe, the New Orleans traders generally made large profits, and many of the most substantial citizens among us, realized their fortunes in this way. The trip from the Upper Ohio to New Orleans, occupied from one to two months, according to the stage of the water, and not unfrequently they were snagged and sunk, or run high and dry upon the shifting sandbars of the Mississippi and Ohio. The labor in time of low water was extremely severe and trying upon the constitutions of those engaged: the boats having sometimes to be literally jumped over the shallow places, by means of levers, sometimes a channel to be dug out of the river bottom sufficiently deep to float them, and every other imaginable device adopted to get their cargoes into port. They coasted along, sometimes doing a retail business at the different landings along the river; but Orleans was the general mart to which they all headed. After selling out there, they sometimes cordelled or hauled back their boats the entire route, at others, they sold their crafts, and either took shipping around by way of the Atlantic ports, or took up their march in companies on foot and horseback, through the Indian country, to their place of departure. Marvellous stories are told of flatboatmen's experience in New Orleans and on the return trip, and there is no doubt but that there was a great deal of foundation for the same, both as regards the city and the travelling. Steamboats and railroads have gradually superseded this mode of transportation, and at this day the broad-horn is an object of curiosity, almost, on the Upper Ohio; as are flatboatmen's yarns a subject for incredulous wonder to the rising generation. An occasional flatboat load still leaves for the Southern country, but for the past ten or fifteen years, flatboating may be pronounced obsolete.

As illustrative of the loose notions of the hunters of that day in regard to the Indians, an incident is related, which we do not recollect of having ever seen in print, but which is as well authenticated as the generality of such stories. It seems that an agreement had been entered into with the Indians that they were to have the exclusive privilege of certain hunting grounds west of the Ohio, choice among which for its abundance of game, was the Stillwater country in what is now Harrison and Carrol counties. This arrangement conflicted with the free and easy notions of the pioneers who had been accustomed to roam at their own sweet will, and marauding expeditions into the Indian country were not of unfrequent occurrence. A party from Washington County, Pa., among which were Solomon and Thomas Eury had penetrated to the Stillwater country, in search of game; and Thomas was shot by the Indians while lying by his camp-fire, his body was covered with a bearskin and his faithful dogs were left at his side as if sentinels over him while sleeping; while the wily Indians were posted around to shoot down the balance of the party as they approached to awaken the sleeper. But by some means their presence was detected by the whites just in time for these latter to save themselves by a precipitate flight, pursued by the whole band of Indians. Nothing was done with the Indians on the ground that Eury righteously met his death while trespassing on their privileges; but some years afterwards, Solomon Eury, the brother of the slain man, happened to be in company with a party of Indians, one of whom, while under the influence of liquor, boasted to him that he was the brave who had killed his brother. The taunt so enraged Solomon, that although in time of peace, he instantly repaired to his house without a word, took down his old rifle running 32 to the pound, dressed himself in full scouting costume, and never stopped until he shot the boasting Indian and six of his comrades. He covered their bodies with leaves and branches where they fell, but the stench attracted attention, the crime was traced to Solomon Eury, he was arrested, taken to Mad River Courthouse and imprisoned, but after a mockery of a trial, acquitted; ostensibly, because the evidence was insufficient, but really, because popular opinion would not admit of his being punished for what every frontiersman felt conscious, he would have clone himself, if similarly circumstanced. This incident gives a pretty faithful idea of the state of feeling at the time and of the general character of the pioneers, in respect to their treatment of and by the Indians.

The piping times of peace which followed the almost annihilation of the Indians by Gen. Wayne, were anything but agreeable to the genius of our hero; nevertheless, unwilling to be idle, he betook himself to learn the carpenter's trade, and bound himself in 1794, as an apprentice to the trade for the period of two years and three months at his old stamping ground, Mercersburg, Pa. Previous to this, he had made a trading trip to New Orleans, in March, 1793, and returned by way of Cuba, through Philadelphia, Chambersburg, &c., to Wellsburg. Even this trip, now of little difficulty, was in those days a matter of very considerable moment and goes to show the habitual restlessness of his disposition. It is not probable that Mr. Gass ever became much of a proficient in the carpenter business, although he points to at least one house in Wellsburg, long known while in the occupancy of Wm. Burgess, as the "*Traveller's Rest*" more recently as the "Yellow Hammer's Nest," said house having got sadly out of repute as well as of repair, in latter days, as a specimen of his handiwork in this line. The house at present belongs to Mr. John Gardner and has very recently been put in good repair, its substantial hewn oak timbers promising long to outlive its architect.

He also had the honor of working on a house for James Buchanan, Sr., the father of President Buchanan, at the foot of Sideling Hill, and saw Gen. Washington, at Carlisle, when he came out with the troops in 1794 to suppress the whisky insurrection. In this war, we believe Patrick had no part; he was too much of a patriot to resist the government; and he loved good old Monongahela too well to enlist against the Whisky Boys, so he remained wisely neutral. He was engaged for a period of six months on the house for Mr. Buchanan, during which time he became well acquainted with "little Jimmy" as he still persists in calling our bachelor President, said "little Jimmy" being ten or twelve years younger than Mr. Gass. Little Jimmy, says Mr. Gass, was then a school boy, rather bright for his years; but showing nothing to particularly distinguish him from thousands of other urchins of his age.

The elder Buchanan was an Irishman who emigrated to this country at an early day, and in York county married a Miss Speer, of a family somewhat distinguished for ability in Pennsylvania. Rev. Mathew Speer a distinguished minister of Carlisle, was a brother of Mrs. Buchanan, and to this family the Gass's were also connected by marriage. From his mother, the President must have inherited his qualities as a statesman, for according to Mr. Gass, the elder Buchanan, was not particularly distinguished among his fellow citizens for any other qualities than thrift and success in making money. He was a merchant and accumulated considerable property by supplying the settlers with iron, salt, &c., in exchange for peltry and hard dollars on pretty much his own terms. Mr. Gass, worked here at his trade with occasional intermissions until May, 1799. At this period, during the presidency of the elder Adams, a prominent speck of war appeared in the horizon, being nothing less than the prospect of a rupture with France under the reign of citizen Genet and his French democracy. This was glorious news for our hero, and throwing down his jack plane and apron he again shouldered his musket and enlisted in the 10th Regiment, American army, under command of Gen., Alex. Hamilton. His services in this war appear to have consisted in a series of marches and counter-marches, among the forts and recruiting stations of Western Pennsylvania, without much glory or personal peril. The winter of 1799, he passed in barracks at Carlisle. From Carlisle he was sent to Harper's Ferry, Va., in June, 1800, and was discharged at Little York, Pa., the French war, which promised so much, winding up most ingloriously.

However, Patrick was not to be cheated out of his full share of military glory, by French or American diplomacy; accordingly the ink that recorded his discharge was hardly dry before he again enlisted in the five years service under Maj. Cass, father of Gen. Lewis Cass, the "hero of the broken sword and stump," of political badinage, and the wise diplomatist of the day, who, celebrated for his antipathy for everything British, has rendered his name memorable in the diplomatic annals of the country; and added the crowning glory by enforcing in 1858, upon the British government the final recognition of the principle for which the war of 1812 was fought, without definite result,—"that the American flag rendered sacred from search or visitation on the high seas by foreign authority, the vessel that bore it."

After claiming the supremacy of the seas for centuries, Great Britain at last relinquished the right of search in May, 1858; when the exercise of the claim by British vessels in the Gulf of Mexico, in the attempt to suppress the slave trade asserted to be carried on with Cuba and the Southern states, aroused a burst of popular indignation, produced energetic measures as well as remonstrances from the government of the United States, and resulted in the full, final and unequivocal, and we will do them the justice to say, handsome renunciation by the British Parliament, of all right or claim to search American vessels on the high seas, unless under treaty stipulations.

By this time, intelligence and merit had brought Mr. Gass, into notice; he was promoted from the ranks, and entrusted with several responsible duties in the way of recruiting and in detecting and arresting deserters. The campaign, however, is barren of incidents of sufficient interest for detail. In 1800, the detachment to which he belonged under Gen. Wilkinson of revolutionary memory, noted for his connection with the "Cabal" and his ignominious defeat in the war of 1812, descended the Ohio in flatboats, passed the Falls on Christmas day, and landed at Wilkinsville, where they wintered in tents and huts. In the Fall of 1801, he went with a company under Capt. Bissell, up Tennessee River, and in the Fall of 1802, the same with a company of artillery were sent to Kaskaskia, Illinois, where they remained until the Fall of 1803, when a call was made for volunteers for the government expedition under Lewis & Clark, being an experimental overland journey across the Rocky mountains, into Oregon Territory.

CHAPTER II.

OVERLAND JOURNEY TO THE PACIFIC

This expedition was projected during the administration of President Jefferson, partly for scientific purposes and partly for the purpose of giving eclat to his administration. The sage of Monticello, the most philosophic of all our presidents, took a just pride in all that related to the literature of the country, and the unexplored fields of the country west of the Mississippi, then not only a barren but an unknown waste, offered a fair opportunity for him not only to gratify his taste and add to his own renown as the patron of such an enterprise, but substantially to add to the material knowledge of the world. With the exception of some trivial contributions made to the stock of general information in regard to this *terra incognita* by the Hudson's Bay Company, who sent out an expedition of discovery under the command of Mr. Hearn, which lasted from December 1770, to June 1772, and explored the country between Churchill river and the mouth of Coppermine between latitude 58 deg., and 72 deg., north, very little authentic information had been recorded. In fact, no regularly organized attempt at exploration for such a purpose, appears to have been made prior to that of Messrs. Lewis and Clark. The expedition of Mr. Hearn, appears to leave been purely of a commercial character, and so far as geographical or scientific objects were concerned, seems to have been barren of results. The individual enterprise and perseverence of the Canada traders, supplied far more general and accurate knowledge of the country. Prior to 1789, they had located trading posts from Canada almost to the Rocky mountains, and about this time they organized themselves together under the general name of the North west company. The hunters and trappers belonging to this company had a tolerably correct practical knowledge of the geography of the country, many of them acquired a knowledge of the dialects of the Indians among whom they traded, and communicated pretty correct ideas of their manners and customs.

In this year 1789, Mr. McKenzie, explored the country between Fort Chippewayen and lake of the Hills, in latitude 58 deg., by the way of Slave river, Slave lake and M'Kenzie river, to the mouth of this latter river, at the North sea in latitude 69 deg.; and in the year 1793, again crossed from Pean river in latitude 56 deg., to the Pacific in latitude 52 deg. north. But these explorations having for their object, principally, the discovery of facilities for extending and prosecuting the fur trade, were necessarily too far to the north to pierce the territories proper of the United States, and it became an object to traverse the country in more Southern latitudes. The southern portion of the continent, reaching up as high as latitude 38 deg., had been for a long time known to the Spanish explorers, consequently, the unexplored country lay between 38 and 52 degrees of north latitude, and between the Mississippi river and the Pacific ocean, embracing an area of about 1000 by 1800 miles. Fabulous stories were in circulation in regard to this portion of the territory. The character of the soil was exagerated. Where it was tillable at all, it was represented as of marvellous fertility, and where it was barren, it was represented as an impassable desert. Those singular formations, the *"Mauvause Terres"* where vast masses of rock tower up, in the desert like artificial erections, were seen by the traders, and what is now known to be only the debris of some mighty natural convulsion, was gravely said to be the ruins of mighty cities—Tadmors of the western wilderness.

The mysterious mirage which so befools the physical eye of the wanderers on these arid plains with tantalizing images of fountains and green pastures, seems to have equally befogged the mental vision of the trappers. Everything in relation to the country was magnified or distorted. The wooly horse had his prototype in their camp-fire narrations. The productions, vegetable, animal and mineral, were half fabulous, and the natives were represented as of prodigious size and extraordinary ferocity. It became extremely hard to sift out and discriminate the few grains of truth from such a mass of fable and falsehood. But the time had come when the reign of the imaginary should give place to that of the real. The genius of progress had decreed that the continent should succumb to the dominion of the white man; and though the gold of California was undreamed of, the balmy climate of the Oregon country, and the fertile fields of the Kansas, unappreciated at that day, she had already waved her wand over the land of the setting sun, and brave and gallant spirits sprang up from the abodes of civilization to do her bidding.

It had become essential to the honor of the country, if not to her profit that these fables should be disproved; and that this magnificent scope of country lying within her domain should be opened up to intelligent possession. That its rivers should be traced to their source, their commercial importance noted, their directions determined, that the qualities of the land, the character of its inhabitants, its vegetation, its animals and minerals should be described, that the face of the country should be defined with accuracy, and the most eligible routes to the Pacific should be ascertained, in short that an accurate and as far as possible faithful transcript of the country should be contributed to the general knowledge of mankind, was the main object of this expedition.

An appropriation for the purpose was made by Congress in the year 1803, and the President empowered to take the necessary measures for its prosecution, in response to a confidential message of January 17th, 1803, recommending such an expedition.

Capt. Merriwether Lewis, of Va., was appointed to the command of the expedition. This appointment was partly owing to family influence, Capt. Lewis being a sister's son of the President, and connected with the influential family of the Lewis's, who were favorites of Washington, and became the recipients through him of large tracts of land in Western Virginia. Gen. Andrew Lewis, the commander of the Virginia forces, at the bloody battle of Point Pleasant, with the Indians, in 1774, a bosom friend of Washington, and a brave and meritorious officer, was a grand uncle of the captain.--He did not owe his appointment, however, altogether to family influence, for he had distinguished himself personally in the Indian campaign, under Gen. Wayne, acid was a man of probity and intelligence, as well as of courage and military ability. He was doubtless a wise selection as the leader of the expedition, and Mr. Gass speaks of him in very high terms of commendation as a gentleman and an officer. He was empowered by the President to select his own men, and chose for his second in command, Lieut. William Clark, a man also reputably connected, and well qualified by previous Indian service for his post. He was a brother of George Rodgers Clark, of Kentucky, afterwards Governor of Missouri, with whom he is sometimes confounded. Capt. Lewis came to Kaskaskia in the fall of 1803, in his search for suitable material for such a corps, and among others who volunteered was Mr. Gass, who happened to be stationed at this post, and to whose adventurous disposition the opportunity presented charms that could not be resisted. To travel where white man had never trod before, was an eminence of venture that rose up mountain high in his imagination, and the danger only dared him to undertake it. Patrick Gass was easily enrolled on the Captain's book, as a member of the party, but Patrick had more difficulty in effecting a release from his military engagements. It so happened that the detachment to which he belonged was about going into cantonment for the winter, and Mr. Gass' accomplishments as a carpenter, joined to his other good qualities, made his immediate commander unwilling to part with him. Accordingly, he raised objection to his leaving, but Patrick was resolute on all occasions, and hard to be balked when he once set his mind upon a purpose.—Ascertaining that

Capt. Lewis was on his way to camp, he went out to meet him on the road, and stating his case with soldier-like directness, the result of the conference was that he was forthwith enrolled in the company of explorers, notwithstanding Capt. Bissell's objections. The selection was not confined to military men, but the call for volunteers was made also to civilians. Among the civilians who volunteered, was Geo. Shannon, a brother of Ex-Governor Shannon, of Ohio, who then resided at Pittsburg, and who accompanied the expedition to its final end, and died some years after, in Kentucky. Several of those who volunteered and were accepted, felt their ardor suddenly cool, when the time came for starting. The immediate prospect of exchanging civilization for barbarism, comfort for hardship and safety for certain peril, with the chance of never returning, proved too much for their philosophy; and to use an expressive term, they backed out.

At the time of starting, the expedition consisted of forty-three men, including officers, privates, and a colored servant of Capt. Clark, named York, who afterwards received his freedom in consideration of his services on the expedition. Some authorities make the number thirty-two, but this is incorrect, as appears from the record in Gass' Journal, made at the time. He has omitted to give a list of the names of the party, but the following taken from Shallus' Chronological Tables, published in Philadelphia, in 1817, may be relied upon as correct, as far as it goes. The company, according to this authority, is as follows:

Captains Lewis and Clark; John Ordway, Nathaniel Pryor, Patrick Gass, Sergeants; William Bratton, John Coulter, John Collin, Pit. Crugatte, Reuben Fields, Joseph

Fields, George Gibson, Silas Goodrich, Hugh Hall, John P. Howard, Baptiste Lapage, Fran. Ladische, Hugh M'Neal, John Potts, John Shields, George Shannon, John B. Thompson, William Werner, Alexander Willard, Richard Windsor, Joseph Whitehouse, Robert Frazier, Peter Wiset, Privates; York, negro man, belonging to Capt. Clark.

In November, 1803, the party made its first move in the direction of the Rocky Mountain country. Leaving Kaskaskia, they proceeded up the Mississippi until they came to the river Du Bois, or Wood river, where they halted for the winter, and occupied their time in preparing boats and making arrangements for a final start up the Missouri the following spring. It is probable that during the long and weary months of a winter spent thus on the confines of civilization, our explorers gave their enterprise many an anxious thought; and it is not improbable that in those hours of comparative inactivity they more than at any subsequent period regretted the enterprise in which they were engaged. There is nothing like constant activity to keep up the courage and the confidence of men, and nothing dissatisfies them sooner with their condition than enforced idleness.—However, they were not entirely unemployed, but found exercise in providing for their subsistence, by hunting, and in preparing boats and in making other arrangements preparatory to the actual commencement of the journey on the opening of spring. Besides this, they had put their hands to the plough, and felt that it would be unmanly and cowardly to look back. Having embarked in an enterprise upon which they felt that the eyes of the nation as well as the attention of the government were bent, they felt that their individual honors were involved, and whatever the hazzard, they could not now think of anything else than prosecuting it to the end.

At last, Monday, the 4th day of May 1804 dawned, bright and pleasant, arguing a successful and safe journey; and elate with high hopes and bright anticipations, and with but a passing thought of regret at leaving the abodes of civilization they started on their perilous journey. They crossed the Mississippi under command of Lieutenant, now Capt. Clarke, Capt. Lewis, being left behind, to overtake them in a few days, and commenced the ascent of the Missouri, the entire expedition being embarked in a Bateau and two Periogues. The little fleet made but slow headway against the rapid current of the river, and by nightfall they had accomplished but six miles up the stream. However a commencement was made, and a after the reflections that usually follow such an event during the first pause, the expedition proceeded with a better heart and a more settled determination. "The determined and resolute character of the corps," says Mr. Gass in his Journal, "and the confidence which pervaded all ranks, dispelled every emotion of fear and anxiety for the present, while a sense of duty and of the honor which would attend the completion of the objects of the expedition; a wish to gratify the expectations of the government and of our fellow citizens, with the feelings which novelty and discovery almost invariably inspire, seemed to ensure us ample support in our future toils, sufferings and dangers."

Day by day they journeyed up the turbid and silent river; on the 16th, they reached the old French village of St. Charles, and as they fired a gun by way of salute, the inhabitants flocked to see them, and on the 21st, being joined by Capt. Lewis, they left the hospitable Frenchmen under a salute of three cheers; which they returned with three more, and three discharges from their guns, and again commenced their toilsome road. By the 25th, they had reached the last white settlement, the small French village of St. Johns, above the mouth of the Wood river, where the river banks were high and the land was rich. Above the mouth of the Gasconade, here 157 yards wide, the party halted, on the 28th, inspected the arms and provisions and sent several men out to hunt, and by the 1st of June, they had reached the mouth of the Osage, here about one fourth the width of the Missouri itself. Their hunters represented the land as the best they had ever seen, and abounding with game. Up the Osage, about 200 miles resided the Osage Indians, a people of large size, well proportioned and very warlike; against any possible collision with whom they thought it prudent to take all reasonable precaution, and in the event of an unfortunate contingency to have themselves in readiness to repel an attack. Their arms and ammunition were accordingly ascertained here to be in good order for any emergency. However, the event showed these precautions unnecessary, for no attempt at interference with them was made by the Indians who seemed indeed universally peaceably disposed. Up to this time they had been without an interpreter--some one through whom they could communicate with the Indians whom they might encounter on their route, but fortunately on the 12th of January they fell in with a party of Sioux on their way to St. Louis with fur and peltry, among whom was an old Frenchman, who professed ability to speak the language of all the Missouri Indians. On the strength of his profession, advantageous offers were made and he was induced to go with the expedition, in the capacity of interpreter, and afterwards proved a most valuable adjunct to the literati of the party, though the sequel showed that in making such extensive professions, he considerably overated his acquaintance with the modern languages. However, necessity, as she knows no laws, must have no scruples; and as the balance of the party were much more ignorant than he, the interpreter was received into the first society the expedition afforded and his gift of tongues duly appreciated.

On the 26th, our voyagers reached the mouth of the Kansas, here 230 yards wide; and as Mr. Gass observes, navigable for a great distance. The intermediate country is described by him as being generally remarkably fertile--a beautiful country, abounding in excellent timber and an abundance of game. Recent events have brought this country into notice and have demonstrated the fidelity of these explorers in their description of the Kansas country, as well as the excellence of their judgement in regard to the qualities of the land.

The navigation of the Missouri was very similar then, to what it is now. At one place we read of their bateau being nearly upset by being caught on a riffle, at another of all hands pulling her against the rapid current by a rope, which broke and nearly caused her loss, then again they pulled around sand bars, and the next thing had to dodge the drift which came down in huge masses. At one time the shores were covered with mulberry trees, in a short time after suitable timber could not be found sufficient to make a pair of oars. An occasional Frenchman would be seen, living solitary and alone, sometimes a stray horse would greet their vision and here and there, they would pass a deserted hut, once occupied by some trapper. The men were sent out to hunt in small parties, sometimes lost themselves in the prairies, and the expedition would have to halt and wait for the stragglers. Deer were frequently killed and their flesh furnished a large portion of the subsistence of the company. Beaver were also plenty, rare birds and animals were of frequent occurrence, specimens of all of which were killed and their skins stuffed for preservation.

By the 4th July, they had reached a point on the Missouri, where Pond Creek enters its waters, and impelled by the spirit of patriotism which seemed to actuate them in all their journeyings, they signalized their appreciation of the day by firing their swivel at daybreak, taking a grand dinner at noon, and christening their encampment *Independence.* The departing day they saluted with another gun. At the feast on the 4th, one of the party was bitten with a snake, that the snake "got into his boots" our author does not state, but considering the time and the circumstances, such an accident was highly excusable, if not probable; at any rate the bite was not dangerous, as he quietly observes.

The glorious 4th, properly celebrated, the voyage was again resumed. Passing a creek called water-which cries, or the weeping stream, they travelled to the 21st, without meeting any incidents of moment, when they reached the mouth of the great river Platte, here, three quarters of a mile wide, and upon whose waters lived numerous tribes of Indians. To these Indians, a deputation was sent to inform them officially of the change in the administration of the U. S. government, and propose a treaty. Their communications and overtures were received with appropriate and becoming gravity, and by the 4th August 1804, proper arrangements were readily effected. The place of conference was called Council Bluffs, by this party. The present "Council Bluffs," in the state of Iowa, although not identical with; is yet in the immediate vicinity of the site. Six of the Indian delegation were here made chiefs, under their "great white father" the President, with which honors they appeared highly pleased.

After this conference was concluded, the party again took up its line of march toward the head waters of the muddy river, their time being variously employed in navigating their crafts, shooting game and fishing, and taking observations of the country. On the 15th, Capt. Clarke and twelve men took 709 fish, among them some catfish of enormous proportions, which proved quite an agreeable addition to their stock of provisions. Here the party experienced the first serious loss that had befallen them, in the death of one of their number, Sergeant Floyd, who was taken sick on the 19th, and died on the 20th. He was the youngest man of the corps, a Kentuckian by birth, and a distant relative of Capt. Clarke. Being naturally of a delicate constitution he had embarked on this expedition in the hope of acquiring better health, but the exposure, superadded to imprudence, was too severe, and he had to succumb in spite of all that could be done to save him. The immediate cause of his death was as follows: He had been amusing himself and carousing at an Indian dance until he became overheated and it being his duty to stand guard that night, he threw himself down on a sand bar of the Missouri, despising the shelter of a tent offered him by his comrade on guard, and was soon seized with the cramp cholic, which terminated his life. During his short illness he received the kindest attentions his comrades could bestow, and his decease was sincerely deplored. But they were not the men to indulge in vain regrets, nor was it a time to indulge in sentimental reflections on the uncertainty of life. They mourned him with a manly sorrow, but his melancholy fate did not deter them from prosecution of their duty. He was buried on the wide prairie, where the desert wild wind sings the requiem of their first to die; and the river over which his spirit broods bears to this day the name of Floyd, given it by his officers in honor of his virtues.

They reached, by the 29th, the country of the far-famed Sioux, whose lodges, to the number of 40, of better material and make than general, were situated about 9 miles from the Missouri, up the river Sacque. Sixty of them came to the camp of the whites, as a peace delegation, and as a token of their sincerity, killed a dog, and treated their white brethren to a dance, in cheap recognition of which, Capt. Lewis constituted five of them chiefs, and presented them with a grained deerskin, to stretch over a keg by way of primitive drum, with which instrument of music, the Indians seemed wonderfully delighted. When their drum was made, a jubilee seems to have been gotten up expressly for the purpose of trying the music that was in it. They all assembled around a couple of fires made for the purpose, and while two of them beat on the drum, a dozen of the rest rattled little bags of dried skin, in which were beads or pebbles, by way of accompaniment, while the dancers, some of them with necklaces of white bear's claws of three inches in length, to the number of twenty or thirty, kept up their performance until "broad daylight in the morning." No squaws, says our author, made their appearance in this dance, whence we conclude that the "stag dance" is not peculiar to the uprorious youths of white blood who occasionally indulge in such exclusive saltatory exercise.

Unfortunately, here, their French interpreter, overcome by the importunities of his Indian friends, left them, having had a better bid from the chiefs of the party, to accompany them to Washington, in the capacity of interpreter for them.

On Sunday, the 2d of September, they encamped opposite an ancient earthern breast-work, 2500 yards in length, running parallel to the Missouri, and with wing walls at right angles, very similar to the Indian fortifications now known to be of frequent occurence in the west.

The question of who were the builders of these works and what is their history has occupied the time and attention of antiquarians for a great many years, but as yet, it is involved in impenetrable mystery. An interesting memoir, by Mr. J. A. Lapham, published under the patronage of the Smithsonian Institute, throws some light on the physical features of these antiquities, which to a remarkable extent, abound in the State of Wisconsin. Under his surveys, the lines as drafted on paper, assume the figures of various animals, deified to this day by the Indians, such as lizzards, turtles, buffalo, &c., a fact which very readily escaped the cursory notice of the earlier travellers, overgrown as were many of the sites with trees and brushwood, but which is material, as going to show that they were intended rather for religious uses, than for purposes of war or defence. This theory is also confirmed by the fact that many of them are elevated only a few inches above the surface of the ground, apparently mere embossments or relievos. At the extreme end of a prairie, 4 ½ miles west of the Mississippi, and the same distance east of the Little St. Francis, exists a curious erection, described as follows, by a correspondent of the St. Louis "Republican": It consists of an oblong square averaging 225 feet each way, with an altitude of twenty seven feet on the southside and twenty one on the north, on the border of what was once a lake, with an area of an acre of level land on the top. The foundation was commenced on a level with the subjacent land, and consisted of a coat of plaster seven inches thick, and burned in several places, on which was placed the dry composition consisting of clay, sand, lime, ashes, pounded shells, and charcoal, carefully mixed, and beat to a hard concrete substance, and so on, until, the height above named was obtained, and then a coat of plastering had been spread over the whole work three inches thick, and burned to a brick redness; but before burning the common wild cane was split and the concave side turned down, and laid longitudinally close together, and pressed into the soft plaster, so that the impressions are now as visible as ever; the whole intermediate space between the two coats of plaster being of the composition above named, in the recesses of which were often found pots inside of which were human skulls, sound and bottom upwards, and other pots sound as ever, full of dry and fresh looking ashes, as though they had been burning incense. It is evident that this large mound was not a place of burial, as no skeletons were found and the adjacent fields are full. Neither do the smaller mounds, contiguous and around the larger one, seem to have been designed for that purpose. Many animal and some human bones were found in the body of the mound, together with images and fragments of ivory, marble and mica."

It has been remarked in this connection that those works are uniformly on what is called the second banks of the rivers, and from this assumption, it is argued that their origin dates back to a period anterior to that when the present channels of the rivers were excavated. This, is not strictly true, and is giving them antiquity unwarranted, at least, by observation among the tumuli of the valley of the Ohio river. These latter are not uniformly, though generally, on the second banks of the river; their location seeming to have been determined on the former, rather by the gravelly character of the material, than by the absence of a first or more alluvial bottom, on which to place them. The existence of even one, on ground of this latter quality, proves incontestably, that the builders lived subsequent to the operation of the causes whatever they were, that produced the second banks of the Ohio. To that period even, the geologists can only approximate in their calculations, and it is giving them a place sufficiently back in remote antiquity, when we say that they were founded not necessarily prior to the formation of the alluvial banks of the western rivers. Remains of this kind are fund in some cases even on the alluvial bottoms of the creeks flowing into the Ohio, as for instance in the neighborhood of Bethany, Brooke County, Va., six miles distant from the river, there were several small ones, now nearly obliterated. One of these was opened by some students a few years ago and found to contain little, if any thing else than a few human bones, giving no evidence that it had been erected for any other purpose than as a monument to the memory of the person buried beneath it. In fact, few of the Indian mounds, that have been explored, have rewarded the labor of their explorers, other, than by convincing them that there was very little to be found. The great mound at Grave Creek, promised some developments but they are considered somewhat apocryphal. A few bones, relics of pottery charred corn, shells, stone implements of war or labor, an occasional scrap of rudely shaped native copper, comprise about all that is generally to be found under these immense heaps of earth, piled, doubtless in barbarian pride, over the remains of some ancient chieftain, to signify by their stupendous size his corresponding importance in their eyes; and by their interior poverty, to warrant them against curiosity or cupidity.

To following account of an antique engraved stone, found some years ago in the Grave Creek mound on the Ohio, has recently attracted attention by the paper of Dr. Wills DeHass, read before the Ethnologic society of New York. This very curious relic of antiquity, as Dr. DeHass appears to have proved it to be, was noticed some years ago by W. B. Hodgson, Esq., of Savannah, in his "Notes on Northern Africa, the Sahara, and Soudan:" Mr. Hodgson, says: "Near one of the skeletons in the lower vault was found the stone in question, with three lines of alphabetic characters.—It is of an oval form, three-fourths of an inch thick, and its material is a fine sand-stone. This is the only example, I believe, of ancient alphabetic inscription in North America. The inscriptions on the Dighton rock and the pictorial writing of Mexico and Yucatan, are symbolic, not alphabetic. The history of this trilinear lapidary inscription, I had at first regarded as apochryphal. Mr. Schoolcraft has, however confirmed it anti described the stone. Who was the gorgeous chieftain whose engraved signet was found by his side? Did he come from the Canary islands, where the Numidian characters and language prevailed? Shall we recur to the lost Atlantis? Could any of the Carthagenian or African vessels, which usually visited the "Fortunate" or Canary islands, have been carried to the New World? The peopling of America is quite as likely to be due to Africa and Europe as to Asia. History preserves the memory of the circumnavigation of Africa by several expeditions. The Periplus of Hanno, the Carthagenian, was the subject of a written narrative. With these historical indications that the Atlantic was in early ages navigated by Mediterranean vessels, I find no difficulty in supposing, the stone in question to have been brought thence."

The fact of huge trees of many hundred years growth upon their ruins, incontestibly establishes a very remote antiquity, and the occasional discovery of relics, displaying some proficiency in the mechanic arts, as certainly proves that their origin is beyond the presence tribes of Indians, who are themselves as much in the dark as to these points, as are the whites themselves. Even their traditions are silent, and unlike the Egyptians, the founders of these monuments have left not even hieroglyphics, which the art and industry of some yankee Champolion or Layard, might peradventure render into readable English. They are impenetrable mysteries, and although they will probably always so remain, they will never cease to be objects of curiosity and research, until under the utilitarian hand of industry, the ruthless plough shall level them with the land, and blot out forever and forever, the little pitiful vestige that remains of what may have been once a mighty, a prosperous and a happy people.

By this time they had come into the prairie country of the Poncas Indians, on the waters of "Rapid-water-river," Plum and White Paint creeks; the diversified nature of the landscape has changed, and instead of the gently rolling plains of the Kansas, the eye wandered over interminable levels, while the river meandered with a more sluggish current between low banks and bluffs more or less high of varied colored clay. The country still continued well timbered, and game in abundance. About this time, Capts. Lewis and Clark, with all the party except the camp-guard, made a foray upon a village of prairie dogs, and though they worked all day and deluged their holes with torrents of water with all the vessels they could extemporize, nightfall found them the possessors of but one unlucky dog, whose points noted in silence, and hide quickly prepared by the naturalist of the Expedition, perhaps figures to this day among the curiosities of Washington City.

Pursuing the tenor of their way, now occasionally diversified with tugging their boats over the frequent shallows of the river, and occasionally adding some rare animal, bird, petrifaction or other curiosity to their collection of novelties, not much of interest occurs in the narration of their journey. By the 20th they had reached a long chain of bluffs, on the north side of the Missouri, of a dark color, the earth of which "dissolves like sugar," and the mixture of large quantities of which in the rapid current, gives its waters their muddy tinge.

On the 25th, another conference took place between the Captains and a delegation of the Teeton branch of the Sioux Indians, which resulted in a mutual exchange of civilities, the making of several of the Indians, chiefs, and came near ending in a brush. This occurred about in this wise: After the ceremonies of the conference were over, Capt. Clarke, sent the new made chiefs ashore in the Periogue, with some of his men, but when they landed, the Indians had taken such a fancy to the boat, that they laid claim to it, and were disposed to prevent its return to its proper owners. To Capt. Clarke's threats they replied that they had soldiers as good as his, and numerous as the leaves of the trees, but whimsically enough, when he told them he had medicine enough in his boat to kill twenty such nations in one day, the magnitude of the idea quite conquered them, and they surrendered the boat in dismay, asserting apologetically, that they only wanted the party to stay with them over night that their women and children might see the boat. So favorable was the impression made upon the simple natives by this medicine talk of Capt. Clarke, that the next we hear of them, eight sturdy savages are carrying Capt. Lewis, and as many more, Capt. Clarke, on their shoulders in Buffalo robes into their Council house, where not less than a dozen dogs were sacrificed and the night passed in carousing, eating and smoking, in honor of their visitors. This time the squaws took part in the dance.

In regard to these Indians, Mr. Gass, makes the following rather dubious mention: "They are the most friendly people I ever saw; but they will pilfer if they have opportunity. They are also very dirty; the water they make use of, is carried in the paunches of the animals they kill, just as they are emptied without being cleaned. They gave us dishes of victuals of various kinds; I had never seen anything like some of these dishes, nor could I tell of what ingredients or how they were made." Patrick's acquaintance with the Indian cuisine was limited, but his stomach was strong, and not to do discourtesy to the hospitality of his hosts, he was no doubt constrained to partake of many a mess that would not so well have suited his tastes among his more dainty feeding friends at home. But a traveller must be a philosopher, and our hero, simply states the facts without giving us any inkling as to his sensations, or indulging in any reflections upon the differences in taste that prevail in different localities. At this camp, they had a continued round of festivities, in which all hands seemed amiably bent upon contributing to the delight of their guests, until when the time came for leaving in the excess of their kindness they siezed the rope and would not allow them to depart. To speed the parting guest, is a maxim of civilized hospitality, that did not seem to be appreciated by the Sioux, and the neglect came nigh being attended with difficulty, for Capt. Lewis, becoming choleric, was just on the point of giving orders to fire on them, when the point was compromised by a carrat of tobacco being given the chiefs, so that they might go in peace. These anecdotes may seem trifiing enough, but they bear the impress of truth, and give a more correct idea of Indian character than pages of labored description could afford. They show the Indian in his true light before communication with the white man had altered their nature.—Impulsive and impressible as children, with little ideas of the rights of property, superstitious to a degree, tickled into good humor by a glittering bauble, or provoked into unreflecting anger by as slight a cause—generous to a friend, exacting to those in their power, relentless to their enemies, brave and cowardly by turns, crafty and yet simple, their character is a tissue of contradictions and yet consistent with itself. At this time they were comparatively unacquainted with the whites, and the native character having fair opportunity to develop itself, perhaps a truer idea of the real western Indian, can be had from the Journal of Mr. Gass, than can be gained from any subsequent source. Since then, they have become indoctrinated with many new ideas by habitual intercourse with white men, as well as corrupted by his vices, so that the Indian of to day is almost another being from the Indian of half century ago.

By the 1st. of October, they had reached the now Du Chien or Dog river, a large tributary of the Missouri, from the south. Above, the course of river was obstructed by sand bars rendering the navigation difficult. A Frenchman, whom they met with, here, informed them that they would not encounter any more Indians until they came into the country of the Rickarees, and accordingly on the 9th having reached a village of this nation, they prepared to hold a council. The village consisted of about sixty lodges, of the construction of which, Mr. Gass gives the following description. "In a circle of a size suited to the dimensions of the intended lodge, they set up sixteen forked posts five or six feet high, and lay poles from one post to another. Against these poles they lean other poles, slanting from the ground, and extending about four inches above the poles: these are to receive the ends of the upper poles, that support the roof. They next setup four large forks, fifteen feet high, and about ten feet apart, in the middle of the area; and poles or beams between these.— The roof poles are then laid on, extending from the lower poles across the beams which rest on the middle forks, of such a length as to leave a hole at the top for a chimney. The whole is then covered with willow branches, except the chimney and a hole below, to pass through. On toe willow branches they lay grass and lastly clay. At the hole below they build a pen about four feet wide and projecting ten feet from the hut; and hang a buffalo skin, at the entrance of the hut for a door. This labour like every other kind is chiefly performed by the squaws. They raise corn, beans and tobacco. Their tobacco is different from any I had before seen: it answers for smoking, but not for chewing. On our return, I crossed from the island to the boat, with two squaws in a buffalo skin stretched on a frame made of boughs, wove together like a crate or basket for that purpose. Captain Lewis and Captain Clarke held a Council with the Indians, and gave them some presents."

   Here they found two Frenchmen living with the Indians, one to interpret and the other to do their trading. A council was held with this nation which ended in an interchange of presents and of amicable protestations; and the party persued their journey among them not only unmolested, but received with marked civility. Mr. Gass, characterizes the Rickarees as the most cleanly Indians he saw on the voyage as well as the most friendly and industrious. A hunting party, which they encountered in their way back to their village, had, says he, twelve buffalo-skin canoes or boats laden with meat and skins; besides some horses that were going down the bank by land. They gave us part of their meat. The party consisted of men, women and children. Shortly after they saw another party of hunters, who asked them to eat, and were very kind and gave them some meat. One of these requested to speak with our young squaw, who for some time hid herself; but at last came out and spoke with him. She then went on shore and talked with him, and gave him a pair of earrings and drops for leave to come with them; and when the horn blew for all hands to come on board, she left them and came to the boat. She shortly afterwards left them and found another hunting party of Rickarees. In the evening, a short time before they encamped, they met with another hunting party of the same tribe. They had a flock of goats, or antelopes, in the river, and killed upwards of forty of them. Captain Lewis, and one of our hunters went out and killed three of the same flock, of more than a hundred."

They pushed onward toward the country of the Mandans and on their way up encountered a couple of Frenchmen who had been hunting in the nation, but were robbed by a party, of their arms, amunition and peltry, and were on their way back very disconsolate. They were glad to be taken aboard of the boats entertaining hopes that they might, though the interference of Captain Lewis's party, regain their property, and being acquainted with the language, their company was quite an acquisition.

They passed in a short time, the place where the Frenchmen had been robbed, but no Indians were ho be seen in the neighborhood except a hunting party of the Sioux, coming down from the Mandan nation, clothed only in *breech clouts*, notwithstanding that the weather had become extremely cold and disagreeable.

This was in the month of October, 1804, and our travellers are far up the Missouri in the country of the Mandans, with the prospect of an early and severe winter before them, the discovery of an Irishman among these Indians is considered an incident worthy of note, as no doubt was the sight of a white skin from any quarter; but passing on, day by day, they pushed farther into the wilderness, until Oct. 27th, their observations showed them that they had travelled 1610 miles from the mouth of the river Dubois, whence they had first embarked. They had averaged scant ten miles per day from the time of their departure, yet theirs was an "original enterprise, and they had progressed as rapidly as the nature of the circumstances would allow. By this time they began to entertain serious thoughts of going into winter quarters and as it was apparent that they were to domicile with the Mandans, it became good policy on their part to make fair weather with their prospective companions. Accordingly, extensive preparations were made for a grand talk, the display accompanyng which was to strike admiration into their hearts. When the principal men from all the villages of the Mandans had assembled, the swivel was fired from the bow of the Captain's boat, and at 11 o'clock the Commanding officers, rigged in appropriate, though tarnished regimentals, took the Chiefs by the hand with becoming ceremony. Capt. Lewis through the interpreter delivered a speech, gave a suit of clothes to each of the head men and some presents of less value for distribution in the villages. As a special mark of consideration, he presented to the united Mandan nation, an Iron Mill, in which to grind their corn. This marvelous liberality quite conquered them, and in token of everlasting friendship, they presented the Captain with 10 bushels of corn, and a deputation from their number volunteered their services to assist him in selecting a suitable site for a winter encampment.

    Whoever has read the romantic adventures of

Capt. John Smith, among the Indians of Virginia, will discover a striking resemblance between his experience as handed down to us by himself and his chroniclers, and that of our voyagers. The same traits seem to have predominated in both instances, and their exercise has been followed with like results. Both found the Indians disposed to be friendly but treacherous; and both found that hospitality abused could be easily converted into deadly enmity. As Hackluyt says of the Virginia Indians: "They are a people gentle, loving, faithful, void of guile, cruel, bloody, destroying whole tribes in their domestic fueds; using base stratagems against their enemies, whom they invited to feasts and killed." In both cases the facile Indian has yielded to the grasping, robust Anglo-Saxon; and but a few more years will elapse ere the Ricarees, the Sioux, the Mandans, and the redmen of every tribe and kindred that yet linger on our borders, will have gone to join the shades of the Powhatans, the Mohegans, the Narragansetts, and the Pequods, in that eternal hunting ground, where alone, they will be secure from the advancing tread and death-distributing knowledge of the white man.

A spot was soon found, surrounded with cotton-wood and suitably situated for an encampment, and on the 2d of November, 1804, they commenced to prepare their winter quarters. They marked out a square, and erected two equal rows of huts, meeting each other at right angles. They designed to enclose the other two sides of the square with pickets. The exterior side of the enclosure presented an elevation of eighteen feet, the inside of about eight and they were made comfortable against the inclemency of the weather, as well as secure against any tricks of their capricious Indian friends.—About the 16th, there came a heavy fall of snow, and they moved, at once, into their unfinished cabins. They were well supplied with provisions, and, all considered, as comfortable as they could expect to be in their situation.

Winter had now set in, in earnest, and our voyagers improved their time in hunting. Taking advantage of the appearance of the Buffalo, which the snows had driven in upon the river bottoms, they killed a great number; in one expedition they and the Indians together, destroyed some fifty. The Indians mounted on horses trained to the business, shot the animals with arrows. In this business they were very expert. Large quantities of meat were laid in at this time, against the time when the increasing severity of the cold would put an end to hunting. This time was not very long delayed. In a few days the weather became so intensely cold as to freeze proof spirits in fifteen minutes. Several of the party were badly frost-bitten, and even the Indians suffered from the same cause. About this time a hunting party of eight Mandan Indians was attacked by the Sioux, one of their number killed, and their horses, &c., taken by the marauders. The facts were reported. to Capt. Clarke, and he and twenty-three men of the party started in pursuit. They tried to induce a party of the Indians to accompany them, but they declined, owing, as they asserted, to the extreme cold weather, and the expedition was, perhaps, wisely, abandoned.

Christmas day was ushered in by a discharge from their swivel, and a round of small arms by the whole corps, the convivial glass was freely passed, and the American flag was hoisted on the ramparts of the little fort, now first christened Fort Mandan, and its appearance, as it

MANDAN INDIANS. — Page 59.

first waved on the breeze, was greeted with another glass very unanimously drank. The balance of the day was devoted to mirth and jolification and the holliday wound up with a general dance in which all hands participated. The precise location of Fort Mandan, as determined by astronomical observation is, 47 deg., 21m., 32s., north latitude, being near the northern bend of the river and distant by their measurement 1610 miles from its mouth. It is called on the maps of this day Fort Clarke, and is still a place of some resort among the traders in those remote regions.

Here appears a chasm in the narrative of Mr. Gass; dating from the 25th December 1804, until the 1st of January 1805, but the subject of discourse where it breaks off and that with which it resumes, are so marvellously alike, that the imagination of the reader needs little aid to enable him to fill up the gap. It is not likely that a party such as ours, after six months assiduous toil, now that the elements had combined to oppose their further progress, would suffer a holliday common to christendom to pass unimproved, especially when they had the society of the Mandan ladies, plenty to eat and something to drink, with which to divert and console themselves. It is highly probable that the interim was appropriately improved, as the introduction to the next chapter, which dates *Tuesday January* 1st, 1805, states that two shots were fired from that same old swivel in honor of the New Year's day, followed by a glass of good old whiskey from Capt. Lewis, and shortly after another from Capt. Clarke, repeated again after noon and doubtless at divers intermediate intervals, from private flasks. This day wound up with a dance in which our hero and Capt. Lewis figured, and with which, "a great number of the natives, men, women and children who came to see us, appeared highly pleased." Mr. Gass, gives but an indifferent account of the Mandan women as regards their personal appearance, habits and behavior, and intimates that chastity was by no means one of their distinguishing virtues. Contrary to the general characteristics of the Indians of the Atlantic country, the conjugal tie seemed to set but lightly upon the natives of the plains; and departures therefrom, were very leniently regarded. Looseness in this regard, seems indeed to be a prevailing characteristic of the western Indians. While among the aborigines of the Atlantic States, continence, was considered a virtue in both sexes and generally practiced, among all the tribes of the Missouri, it was but little regarded, and adultery and prostitution hardly considered as venial offences. Public opinion and custom however, generally regulate these things even in civilized countries, and it would be uncharitable to apply to the Mandan Indians the same standard of morals that is recognised among people more advanced in civilization. They would be doing as much as could be expected of them, and more than the whites often do, if they did not transgress their own customs, usages, and laws. This, we have no reason to believe they did. The Mandans are described as having lighter complexions than most other Indians, many of the children having light colored or flaxen hair, the children of other tribes being uniformly black haired from birth. During the months of January and February, nothing of much interest occurs, the party being principally occupied in hunting and in the ordinary routine of camp life. Some of the experiences of the hunters were rather disagreeable, owing to the extreme cold, and occasionally to the necessity of partaking of wolf meat, when no better game could be found; but generally, game was abundant, and the blacksmith of the part driving a thriving trade by furnishing the natives with hatchets and other articles of iron for corn, at price that would Blake our modern speculators stare, they were amply supplied with the essentials of comfortable life.

The tedium of this mode of living, was interrupted on the 14th, by the return to camp of a party of four of their men, who, having gone out with four horses for the purpose of bringing some meat that had been stored at a distance from the camp, and had been beset by a party of Sioux Indians, and robbed of three of their horses, came in, and reported the facts to their comrades. The little cantonment was instantly in a ferment, and at midnight, Capt. Lewis having called for volunteers, twenty of the party promptly turned out to chastise the marauders and recover the horses. The promptness of the response, is an evidence of the spirit that actuated the entire party. By the 16th, having accomplished some fifty miles of a march, in the pursuit, they came upon a deserted camp, which lead been burned by the Indians and which was still smoking; but the savages had dispersed and fled into the plains and further pursuit was deemed useless. The bold demeanor of the whites, had impressed the Indians with becoming degree of awe, and although they boasted of their robbery and sent word that the Sioux intended to massacre the entire party in the spring, they were not much terrified by the threat nor did they afterwards meet with any serious trouble from these doughty adversaries.—Some of the Indians in this neighborhood manifested a mischievous spirit. The party concealed some meat near some deserted huts, which was found and destroyed, and the huts burned by them, but beyond such petty annoyances they experienced little trouble from the Indians during their winter's residence among them. Mr. Gass, speaks of a beautiful breed of white rabbits that abounded in that section, and also informs us how the Indians managed to keep their horses in winter.—They had a great many of these, and during the daytime they suffered them to browse around and subsist the best they could; at night they introduced them into their huts and fed them upon *cotton wood branches*. Upon this meagre fodder they kept in tolerable condition and were serviceable until the return of grass when they fared more sumptuously.

It was now determined to send a portion of the party, with what skins and other specimens they had procured and an account of their proceedings to date, down to St. Louis, while the balance of the party, pursued their journey up the Missouri. Accordingly, all hands busied themselves with the task of preparing canoes, which by the way had to be carried a mile and a half to water before they could be launched. Six canoes were soon finished, but on trial, their capacity was found insufficient, and it was determined to send the large periogue back, with the returning party.

On Sunday, the 7th March, they broke up their encampment, thirty-one men and a woman going up the river with two periogues; and six canoes and thirteen going down with the large periogue, in which were packed the curiosities, "Buffalo robes and horns of the mountain ram of prodigious size for the President."

The woman mentioned in the preceding paragraph, was one of the wives of their interpreter, whose presence was expected to be of benefit to them in their passage through the Snake Indians, to which tribe she belonged. The river still continued easily navigable; and they made good headway, although they had now reached a point higher up, than had ever before been attained by white men. The character of the country began to change, indications of volcanic action became of frequent occurrence, and the hills seemed sterile and naked of any appearance of vegetation, though there still appeared to be plenty of game of various kinds.—A new article of diet here appears to have come in vogue, nothing less than wild geese eggs, which they found deposited by those usually stupid birds in nests high up in the trees, and another seasonal delicacy was found in the young Buffalo calves, a number of which were about this time killed by the hunters of the party. Our journalist here remarks as a "singular circumstance" what others have since found out to their sorrow, that in this region there is no dew and very little rain, and with an astuteness worthy of Professor Espy, he enquires whether it can be owing to want of timber. They had now got upon the borders of the desert country known as the "plains" by later travellers, in the overland journey to Oregon and California, and which proves so disastrous from almost utter want of water and scarcity of grass for stock.

This was at the mouth of Yellow Stone river, which they ascertained by measurement to be, in width, 297 yards of water and 561 of sand, while the Missouri was 331 yards of water and 190 of beach, the current of the latter river continuing strong, while that of the former is sluggish and shallow. This point is given at 1888 miles above the mouth of the Missouri, and 278 from their winter quarters at Fort Mandan. Portions of the country in this neighborhood are represented as very fertile, though indications began to multiply of their near approach to the Rocky mountains. They made an unsuccessful attempt here to kill some mountain sheep being the first they had yet seen, but though they failed in this, Capt. Lewis succeeded in dispatching another stranger with whose kindred they were destined to become better acquainted, being nothing less than a huge "grizzly."

The 1st of May, when the "cotton wood leaves were as large as dollars," they were greeted with a snow storm which compelled the boats to lay up; during which time the hunters killed several deer, and some of them discovered some red cloth in an old Indian camp, which it was thought had been offered up by the Indians by way of sacrifice to their deity—"the Indians," says Mr. Gass, "having some knowledge of a Supreme Being, and this, being their mode of worship."

It is a singular fact that not a single tribe of Indians has been discovered in North America, but has had some idea of the existence of a Supreme Being, and the immortality of the soul. There is a striking similarity in their beliefs and superstitions in this respect; and notwithstanding their general ignorance, their views are more philosophical than those of other nations much more advanced in civilization. They had better ideas of natural religion than had the Greeks and Romans, notwithstanding the fact, that these latter were the most polished, as well as the most intellectually acute nations of antiquity. The conception of one Great Author of all, to whom all are forever responsible, was the central idea upon which all the minor beliefs were founded; and though many of their superstitious notions appear to us whimsical and absurd, yet this grand central idea may be discerned, more or less distinctly, through all. They had a vague notion of the truth, handed down from generation to generation and though cumbered and distorted with error, their minds appeared to grope in the dark in the vain effort to reach the light without divine revelation. That they came so near it, is more to be wondered at, than that they should be lost in the mists of the ignorance that beset them.

The hills which for many days had been barren of vegetation, now began to make a sparse display of pine and cedar trees, the verdure of which was quite enlivening to the spirits of our voyagers, while the surrounding scenery wore the appearance of architectural ruins noticed by travellers as the *mauvais terres,* or bad lands, though the river continued wide and in handsome order for navigation. Bears become more and more abundant and bear meat the staple article of their cookery. A large brown bear being wounded by six of the party, made battle and came near making specimens of his assailants; but powder and lead, backed by overwhelming numbers, were too much for him, and after a desperate fight he had to succumb. The natives, stood in dread of these grizzly gentleman,—not a few of their braves having fallen victims to their prowess in hand to hand encounters. The mountain sheep also become more abundant as they approach the mountain and they at last succeed in killing some of them. They are described as more resembling the ibex than the sheep, being covered with a long soft hair of a dun color instead of wool, and resembling sheep only in the head, horns and feet. The horns of one specimen were two feet long and four inches in circumference at the base. In size, they are something larger than the deer. It differs from the deer in never shedding its horns. Naturalists have since assigned them a place in the family of the ruminantia. They also, killed a large brown bear of the following formidable dimensions:--3 feet 5 in. around the head, 3 feet 11 inches around the neck, 6 feet around the breast; the length 8 feet 7 inches, around middle of the forelegs 23 inches, and with talons 4½ inches in length, and sharp as needles. Such customers as this were becoming uncomfortably abundant, and their company was not particularly sought by the hunters to whom they sometimes gave chase. These bears are very tenacious of life and when pressed to desperation are particularly dangerous. The natives very seldom dared to attack them, having lost many of their braves in such encounters. The bears themselves, are not afraid of man, but will attack him without hesitation, and even when well armed the struggle is one of considerable risk to the hunter. Some of the exploring party discovered a large brown bear about this time at the mouth of a small creek, and a party of six men sallied out to kill it. The bear, took refuge in a thicket, and stood at bay growling terribly. The men advanced, and three of them fired simultaneously, aiming at the most vital parts of his body; but although riddled by their balls, he only seemed roused to fury. Rushing from his covert, the remaining three barely had time to discharge their pieces before he was among them, and the issue came near being a tragical one. Two of the men were badly torn by the claws of the infuriated beast before he could be dispatched, which was, however, finally effected after a desperate fight. His carcass weighed in the neighborhood of eleven hundred pounds.

Having now, Monday, 27th May 1805, fairly entered upon a country presenting nothing but barreness and desolation to the eye; and traversed a distance of 2300 miles; our journalist proceeds to give a brief recapitulation in regard to the topography of the country, which we can do no better than to give in his own words.

"From the mouth of the Missouri to the river Platte, a distance of more than 600 miles, the land is generally of a good quality, with a sufficient quantity of timber; in many places very rich and the country pleasant and agreeable. From the confluence of the river Platte with the Missouri in the sterile desert we lately entered, a distance of upwards of 1500 miles, the soil is less rich, and except in the bottoms the land is of inferior quality, but may in general be called good second rate land. The country is rather hilly than level though not mountainous, rocky or stony. The hills in their unsheltered state are much exposed to be washed by heavy rains. This kind of country and soil which has fallen under our observation in our journey up the Missouri, extends, it is understood, to a great distance on both sides of the river. Along the Missouri and the waters which flow into it, cotton wood and willows are frequent in the bottoms and islands; but the upland is almost entirely without timber, and consists of large prairies and plains whose boundary, the eye cannot reach. The grass is generally short on these immense natural pastures, which in the proper season are decorated with blossoms and flowers of various colors. The views from these hills are interesting and grand. Wide extended plains, with their hills and vales, stretching away in lessening wavy ridges, until by their distance they fade from sight; large rivers and streams in their rapid course winding in various meanders; groves of cotton wood and willow along the waters intersecting the landscape in different directions, dividing them into various forms, at length appearing like dark clouds, and sinking in the horizon; these enlivened with the buffalo, elk, deer, and other animals, which in vast numbers feed upon the plains, or pursue their prey, are the prominent objects which compose the extensive prospects presented to the view, and strike the attention of the beholder."

The Missouri like all our western rivers is diversified with wooded islands, in general not so large, nor so picturesque however as those which gain for the Ohio its appellation of the beautiful river. Along its course, were Indian paths, and roads made by the Buffalo and other animals often ten feet in width and equalling in directness from point to point, roads made by human engineers; if indeed in many cases these lines made by nature's engineers do not excel those laid down by compass and chain. They had now come into a savage and sterile country with nothing to relieve its wild monotony, save the roaring of the waters, as they whirled and dashed among the rocks and the increased caution and greater labor which the growing difficulty of the navigation rendered imperative. While tied up to the shore at night, a Buffalo swimming the river chanced to land at the periogue; and making a flying leap to clear the obstruction he landed in the boat, nearly capsizing it and clashing among the men, who were sleeping, they awoke in great consternation thinking that the devil had come in person to torment them for their sins. He gave them a serious fright but did no injury, save disabling a couple of guns that lay in his way. At one place here, the Indians had killed over a hundred of these animals in one drove by driving them over a precipice, which seemed to be a favorite mode among these Indians. Some of the appearances, here, were striking in the extreme, high walls of solid rock, stood up grey and perpendicular, 200 feet in altitude, by eight or ten in thickness, and of interminable length, occasionally, a column erected itself in solitary grandeur, like the chimney-stack of some crumbled down manufactory; long ranges of shattered ruins appeared as though the hand of time had been playing havoc with a deserted city; and the whole scenery had that wierd and melancholy aspect, which according well with the brooding and gloomy silence so suggestive of the world to come, so haunts, with visions of the supernatural and infernal, minds prone to superstition.

June 3d, 1805, the commanding officers being in a quandary which of two forks of about equal size it was proper to ascend, a couple of parties were detailed to try each, and thus determine which was the Missouri proper. Two days were passed in reconoitering, and finally the south branch was reported the best for navigation, and up it they went, for the distance of sixty miles. The other branch is called Maria's river. The decreasing volume of the river admonished them to leave a part of their luggage and stores; and accordingly a portion of the party busied themselves in preparing a hole in which to bury their surplus pork, corn, powder, lead, &c., to the amount of a thousand weight; while Capt. Lewis took a meridian observation in order to determine their precise location. The result was 47 deg., 24 m., 12 s., North latitude. At the mouth of Maria's river the large periogue was concealed under heaps of brush, and near by was deposited their surplus stores, to be ready for them on their return.

On the morning of the 12th, they left this point and proceeded up the river, encountering great difficulty, owing to the numerous rapids; sometimes having to take the boats entirely out of water and transport them by land, on wheels extemporized for the purpose, and by the 18th, so toilsome had their progress become, that it was determined to bury more of their luggage, which was accordingly done. The boats were mounted on wheels, and the party accompanied it across a prairie, to the next point of embarkation, 16 miles distant, which consumed the day. Up to the Fort Mandan, the general direction of the river had been Northwest, thence to this point, nearly due West, but shortly before reaching this point, now called Clark's Falls, the course of the river turns to the South, and traverses some 200 miles almost directly to the South. The Falls or Rapids, are in the aggregate 362 feet in height, and extend for eighteen miles. After passing them, the character of the river as well as of the country changes, the river is smoother and more equable in its current, while the country appears to be more level, with mountains covered with snow, surrounding them in the distance.—One of the party here had a narrow escape from being devoured by bears. He was attacked by three brown bears, and to save his life, was forced to leap over a precipice, severely injuring himself and breaking his gun.—Another man about the same time, was attacked by a huge he bear, when separated some 200 yards from the balance of the party. His gun, unluckily, missed fire, and the bank was so steep that his companions could not reach him; however, they fired at the bear from a distance, which had the effect of frightening him off, and thus saving their comrade from an ugly encounter, in which the principal risk would have been on his side. Buffalo, elk and deer, as well as bears, appeared to be very plenty in this vicinity, and quite a number were killed by the party. An experiment of covering the iron frame of a boat, which they had brought with them for the purpose, with skins, so as to be easily portable, proved a failure, owing to the impossibility of making it water tight, and they were compelled again to make new canoes, or leave more of their baggage. A couple of canoes were soon constructed and the party pushed on, the plains were covered with a short grass, and the hills from 600 to 1200 feet in altitude almost solid rock, bare of vegetation and seeming to be a favorite haunt of the Rocky Mountain sheep which were seen in great numbers on the very summits. Along the course of the river there was a fringe of cotton wood and bushes, in which a great many deer and other animals were found, and also a great variety of seasonable berries, among which is mentioned the service berry, the choke berry and as particularly large and fine, the black currant. Indians had become extremely scarce and although parties were sent out expressly to find them, they did not discover a native for weeks in succession. A smoke was discovered in the distance, which on investigation proved to rise from an Indian hunting camp, the proprietors of which, evidently taking the whites for enemies had fled into the wilderness. A pole which Mr. Gass had erected as a mark for a small party left behind, and in a cleft of which he had placed a note was knawed down by a beaver and dragged off, giving, the party for whose benefit it was intended, a wearisome tramp of several miles in the wrong direction before their error was detected. A kind of red clay was noted as occurring here, which their squaw informed them was used for war paint.

About this time a singular accident occurred thus related by Mr. Gass. It appears, that some of the party had discovered a fine sulphur spring, which Captain Clarke, the Interpreter, his squaw and child went to look at. During their visit a sudden storm came up, forcing them to shelter under a bank at the mouth of a run. In five minutes time, such was the violence of the torrent, there were seven feet of water in the run and all hands came near being washed away. As it was, they lost a gun, umbrella and surveyor's compass, and barely escaped with their lives. At this place they had another encounter with a bear. On the 30th July 1805, they reached what Mr. Gass calls the Forks, and on the 9th August, the commanding Officers came to the conclusion that these forks might be properly considered the end of the Missouri, and proceeded to name them, Jefferson, Madison and Gallatin, being respectively the north, middle and south branches, coming in nearly at the same spot. The latitude of this confluence of the waters was determined to be about 45 deg., 3 min. north.

We here make an extract from the Journal: "There is very little difference in the size of the three branches. On the bank of the north branch we found a note Captain Clarke had left, informing us he was ahead and had gone up that branch. We went on to the point; and as the men were much fatigued, encamped in order to rest a day or two. After we halted here, it began to rain and continued three hours. About 12 o'clock Capt. Clarke and his men came to our encampment, and told us they had been up both branches a considerable distance, but could discover none of the natives. "There is a beautiful valley at these forks, and a good deal of timber on the branches, chiefly cotton-wood.—Also currants, goose and service berries, and choak cherries on the banks. The deer are plenty too, some of the men went out and killed several to-day." Capt. Clarke, who had been taken sick on the route, is here reported convalescent, and Capt. Lewis, who had traveled ahead of the party, was obliged to camp out by himself in the howling wilderness. He, however, turned up all right in the morning, and the party dividing, Capt. Clarke would explore one branch with his corps and Capt. Lewis, another, with his, leaving notes at prominent places to direct each other in their explorations. The travelling had become difficult and the routes extremely mixed, rendering their progress very slow. At this place, our squaw informed us, "she had been taken prisoner by the Grossventers four or five years ago. From this valley we can discover a large mountain with snow on it, towards the southwest; and expect to pass by the northwest end of it. Capt. Lewis had a meridian altitude here, which gave 45 deg., 22m, 35s. north latitude."

Ascending the north or Jefferson branch, it also forked into Wisdom and Philanthropy branches, the middle one retaining the name of Jefferson, up which they continued. It has been remarked that there is nothing in a name, and that a rose by any other name would smell as sweet; but our explorers seem to have taxed their ingenuity to considerable extent in order to find suitable names for the streams which it was their fortune to christen. It is perhaps to be regretted that they in common with all our other explorers, did not adopt the Indian names of the streams and points, or at least modify them so as to sound properly to English ears. The multiplication of English proper names as applied to geographical objects, is the source of great inconvenience and confusion; while the adoption of many common and vulgar English words, as appellatives, is often in decidedly bad taste. The appellatives of the Indians, generally abounded in vowel sounds, and what was more, had mostly some reference to peculiarities of the thing they designated. The idea was often as poetical as the sound was euphonious. It is to be regretted that our maps do not contain more of them. What can be more suggestive than MIN-NE-HA-HA, the *water which laughs,* as applied to the rippling waters of the MISSI-SIPPI, *father of waters,* smiling back the sunlight from its pebbly bed before the *muddy* MISS-OURI throws its sombre shade on the now sullen and turbulent current. It is evident that the philosophy of names gained nothing in its poetic department from the hard practical sense of Anglo American discoverers. They were inferior in perception of the beautiful and the grand, to the French and infinitely behind the poor Indian whose—"untutored mind, Sees God in the storm and hears him in the wind."

Journeying on, they passed an island which, as it was distant about 3000 miles from their starting point, they christened with some propriety, *"Three thousand mile island;"* the river being only about twenty yards wide and a foot and a half deep, meandering and winding along through the bushes, and frequently compelling the men to wade and drag the canoes through the water which had become icy cold and very disagreeable.—The black or mountain trout of large size abounded, as well as a variety of other fish; while deer and game generally had become scarce, owing to the scant herbage. At the entrance to a gorge in the mountains here, two huge pillars of rock erect themselves like watch towers, guarding the entrance as if to some land of promise, picturesque enough the reader will say, but not say our explorers, realizing the promise of fertility so jealously guarded. From this point to the head waters of the Columbia river, emptying into the Pacific, was but about forty miles. Capt. Lewis had passed over the ground, and returned, bringing with him some twenty of the Snake Indians whom he had encountered, with a like number of horses, with which animals, these Indians fortunately enough were well provided. Mr. Gass, speaks here of the great quantity of service berries found in this region, which seem almost as if providentially provided for the sustenance of such living animals as may happen among these inhospitable gorges of the mountains.

The party now provided with Indian guides followed up the gorge of Jefferson Fork, now a mere mountain brook, until the 19th August, 1805, when they reached the head spring, distant only one mile from the head of one branch of the Columbia.

On the head waters of the Columbia, Indians became more numerous; and on the 20th they came to a village of twenty-five lodges made of willow bark. This was a village of the "Snakes." "They are, "says Mr. Gass, "the poorest and most miserable nation I ever beheld; having scarsely anything to subsist on, except berries and a few fish, which they contrive by some means to take. They have a great many fine horses, and nothing else; and on account of these they are harrassed by other nations. The usual mode of salutation, among the Snakes, is by putting the arms around the neck of the person they wish to salute, in token of friendship." This method, it will be allowed, was more affectionate than agreeable to persons of weak stomachs.

The western Indians, seem generally, to have been more disposed toward a pastoral life than were those east of the Mississippi; and manifested more of a disposition to possess themselves of flocks and herds.—Horses, were found by this party, among all the tribes through which they passed; and often in localities the most unfavorable. They were used for purposes of travel and the chase. In case of emergency their flesh was eaten. The origin of these horses among the Indians is conjectural. They probably descended from Spanish stock imported at the time of the conquest, or subsequent settlements. In the genial climate of the tropics, it is not impossible that a few individuals escaping or turned out into the vast natural pastures, increased and multiplied into the immense herds that appeared over the boundless plains of Mexico and South America. Another very probable source was the Canadas. It is certain that at the discovery of the country the Indians had no idea of horses, in fact, those used by Cortez were actually worshipped by the Mexican Indians, and considered superior beings. In later times the western Indians have still farther devoted themselves to stock raising, and at this date, the Navajos are reported as the possessors of a half million sheep.

These Indians, gave the party a very unfavorable account of the navigation of the Columbia, so much so, indeed, as to induce them to abandon their canoes and undertake the journey by land. Accordingly, they purchased a stock of horses from the Indians, at an average of about

$3 per head in merchandise, and leaving Capt. Clarke, to bring the baggage by water, they continued down the Columbia. The representations of the Indians proved correct, for the river bottom was narrow and the route almost impracticable either by land or water; nevertheless, they persevered—the rocks in some places breast high and no path or trail of any kind to direct them until brought to a dead halt, at a point where "the water was so rapid, and the bed of the river so rocky that going by water appeared impracticable; and the mountains so amazing high, steep and rocky, that it seemed impossible to go along the river by land." Their trials now fairly commenced. The Journal proceeds: "Captain Clarke and our party proceeded down the river with our guide, through a valley about four miles wide, of a rich soil, but almost without timber. There are high mountains on both sides, with some pine trees on them. We went about eight miles and encamped at a fine spring. One of our men remained behind at the village to buy a horse, and did not join us this evening. Five of the Indians came and stayed with us during the night. They told us that they were sometimes reduced to such want, as to be obliged to eat their horses.

Next morning, we began our journey at 7 o'clock and having travelled about a mile, crossed a branch of the river. Here the mountains came so close on the river, we could not get through the narrows, and had to cross a very high mountain about three miles over, and then struck the river again, where there is a small bottom and one lodge of the natives in it, gathering berries, haws and cherries for winter food. We soon had to ascend another large mountain, and had to proceed in the same way until we crossed four of them, when we came to a large creek, where there is a small bottom and three lodges of Indians. Three of our men having gone through the bottom to hunt, came first upon the lodges which greatly alarmed the unhappy natives, who all fell a weeping and began to run off; but the party coming up with the guide relieved them from their fears. They then received us kindly and gave us berries and fish to eat. We remained with them about two hours and gave them some presents. Those of the natives, who are detached in small parties, appear to live better, and to have a larger supply of provisions, than those who live in large villages. The people of these three lodges have gathered a quantity of sun-flower seed, and also, of lambs-quarter, which they pound and mix with service berries, and make of the composition a kind of bread; which appears capable of sustaining life for sometime. On this bread and the fish they take out of the river, these people, who appear to be the most wretched of the human species, chiefly subsist. They gave us some dried salmon, and we proceeded down the river; but with a great deal of difficulty: the mountains being, so close, steep and rocky. The river here is about 80 yards wide, and continually rapid, but not deep. We went about fifteen miles today, and encamped on a small island, as there was no other level place near. Game is scarce, and we killed nothing since the 13th but one deer; and our stock of provisions is exhausted.

"On the 23d, two of the hunters went in search of a buck that had been wounded during the day, and the rest staid in the camp to fish. In the afternoon the men came in from hunting the wounded deer, but could not find him. They killed three prairie hens, or pheasants. We caught some small fish in the night. The natives take their fish by spearing them; their spears for this purpose are poles with bones fixed to the ends of them, with which they strike the fish. They have but four guns in the nation, and catch goats and some other animals by running them down with horses. The dresses of the women are a kind of shifts made of the skins of these goats and mountain sheep, which come down to the middle of the leg. Some of them have robes, but others none. Some of the men have shirts and some are without any. Some also have robes made of beaver and buffalo skins; but there are few of the former. I saw one made of ground hog skins.

"The river at this place is so confined by the mountains that it is not more than twenty yards wide, and very rapid. The mountains on the sides are not less than one thousand feet high and very steep. There are a few pines growing on them. We caught some small fish to day, and our hunters killed five prairie fowls.—These were all we had to subsist on. At 1 o'clock Capt. Clarke and his party returned, after having been down the river about 12 miles. They found it was not possible to go down either by land or water, without much risk and trouble. The water is so rapid and the bed of the river so rocky, that going by water appeared impracticable; and the mountains so amazingly high, steep and rocky, that it seemed impossible to go along the river by land. Our guide speaks of a way to sea, by going up the south fork of this river, getting on to the mountains that way, and then turning to the south west again. Capt. Clarke, therefore wrote a letter to Capt. Lewis, and dispatched a man on horseback to meet him; and we all turned back up the river again, poor and uncomfortable enough, as we had nothing to eat, and there is no game. We proceeded up about three miles, and supperless went to rest for the night.

"Next morning we set out early and had a fine day; passed the Indian camp, where they gave us a little dried salmon, and proceeded back again over the mountains. Some hunters went on ahead and encamped in the valley. Two men went to hunt, and all the rest to fish. We soon caught as many small fish as made with two salmon our guide got from some Indians, a comfortable supper. At dark our hunters came in and had killed but one beaver.

"Monday 26th, we had again a pleasant morning and four hunters went on early ahead, and one man to look for the horses. We breakfasted on the beaver and a salmon, which had been saved from supper the preceding evening. The man who had gone for the horses, having returned without finding them, four or live more went out, and our guide immediately found them. We then, about ten o'clock, proceeded on to the forks, where we found our hunters; but they had killed nothing. So we went up to a small village of the natives, got some fish from them, and lodged there all night.

"Next morning eight of us went out to hunt. I observed some flax growing in the bottoms on this river, but saw no clover or timothy, as I had seen on the Missouri and Jefferson river. There is a kind of wild sage or hyssop, as high as a man's head, full of branches and leaves, which grows in these bottoms, with shrubs of different kinds. In the evening we all came in again and had killed nothing but a fish. We got some more from the natives, which we subsisted on. We lodged here again all night, but heard nothing from Captain Lewis.

On the morning of the 28th of August, I went on to the upper village, where I found Capt. Lewis and his party buying horses. They had got twenty-three, which with two we had, made in the whole twenty-five. I then returned to our camp, a distance of fifteen miles, and arrived there late. I found the weather very cold for the season."

The ascent of the Missouri had been plain sailing in comparison to the navigation of the Columbia, where precipices of a thousand feet elevation came sheer down to the waters edge, so steep that nothing save the venturesome feet of the mountain goat dared to scale them, and of such interminable length, that the most experienced guides were at fault as to expedients to go around or across; while the tumbling torrents at their base as they pitched and dashed over high masses and ledges of rock, bid defiance to any craft that might be constructed to navigate them. They were now in a dilemma, and to add to their troubles, provisions became exhausted and there was very little game with which to replenish. All the skill of their hunters could not keep them supplied with meat, and more than once they were forced to subsist on horse flesh, beaver and dog, with bread made in Indian fashion, of sun-flower seed, lambsquarter, service berries, and dried salmon pounded and incorporated together. This latter article was not so unpalatable, and proved an excellent substitute, now that their scant supply of flour was exhausted. The natives are represented as miserable in the extreme, almost starving, and nearly naked; depending chiefly for subsistence upon fish, which they speared with long poles pointed with sharp and barbed bones for the purpose.

Salmon in the Columbia were abundant and very fine, and well enough it was so, or our voyagers, would have starved to death in a wilderness as inhospitable, otherwise, as the icy deserts of the north, that have proven the burial places of so many gallant and venturesome men. This brings us up to September 1st. 1805; for the past few days our party has encountered difficulties that were almost insurmountable and endured hardships in almost every shape and form with a perseverance that excites our admiration and with a fortitude that should rank them among the foremost travellers of any age.—Though all this narrative of trial, deprivation and hardship, we look in vain in Mr. Gass's journal for a single instance of repining; no regrets sully its pages and no complaints either by him, or by any of his companions. About this time the commanding officers of the part seem to have had a high appreciation of the merits of Mr. Gass, mention being frequently made, in a modest way, of his participation in important services. There is, however, no ostentation about the narrative, all goes along in a smooth matter of fact way, as if the incidents narrated, were of every day occurrence and entitled to no particular mention.

The men bore their hardships manfully and obeyed with unflagging energy and undeviating fidelity the commands of their officers; who, themselves, seem to have been well worthy of the men over whom they were placed. But what less could be expected. The American is a man every inch of him, whether in civil or in military life, whether in command or in subordination. As a soldier he knows his place and his rights as a free man; and the true officer will exact nothing but what he knows will be done with a will; and the man will execute whatever is to be done with a zeal and an intelligence that no other nation can attain. It is this characteristic that makes them invincible as soldiers and renders them notorious for indomitable will, steady perseverence and great achievements in whatever enterprises they engage, either of peace or war. For the next few days they passed through the same difficulties, striving with almost superhuman energy to surmount the last barrier that nature has erected between the opposing surges of the Pacific and the Atlantic, stretching like a huge back-bone the length of the continent and parting the fresh fallen waters of the East and the West to the right and to the left with its adamantine vertebrae. Gradually, however, their course became somewhat smoother, traversing occasional small vallies, like oases, of rich black soil, abounding with herbs, berries and edible roots, and inhabited by the Flathead Indians, who Mr. Gass, denominates the whitest Indians he ever saw, and who are much better provided with worldly geer than their neighbors, the Snakes, from whose country they are about emerging. "The Indian dogs are so hungry and ravenous," says he, "they ate five pair of our mockasons last night. We remained here all day, and recruited our horses to forty and three colts; and made four or five of this nation of Indians chiefs. They are a very friendly people; have plenty of robes and skins for covering, and a large stock of horses, some of which are very good; but they have nothing to eat, but berries, roots and such articles of food. This band is on its way over to the Missouri or Yellow-stone river to hunt buffalo. Next morning we exchanged some of our horses, that were fatigued, with the natives. Four hunters who had been out, killed nothing; we therefore supped upon a small quantity of corn we had yet left. Next day one of the hunters killed two deer; which was a subject of much joy and congratulation. Here we remained to dine, and some rain fell. On the south of this place there are very high mountains covered with snow and timber, and to the north prairie hills. After staying here two hours, we proceeded on down the creek; found the country much the same as that we had passed through in the forenoon; and having travelled about twenty miles since the morning, encamped for the night—having killed two cranes on our way."

On the 9th they pass the mouth of the river of the Flatheads, here 100 yards wide, which they name Clarke's river, and by the 13th they came to a beautiful warm spring, with numerous paths diverging from it, and the waters of which were hot above blood heat. Four of the best hunters of the party, well mounted were out all day as a special party, but neither they nor the balance of the party had much success as the common larder showed but three pitiful pheasants, with which to feed thirty hungry men. In this strait, they resorted; to some portable soup, which was brought along to be used in case of necessity, and so give it body, killed and roasted a colt, which our hero says, made it "good eating." For some days after this, the prevailing diet was portable soup, parched corn and roasted colt, with no particular complaint except as to quantity. A horse fell over a precipice of a hundred feet, without being much hurt as Mr. Gass observes, owing to the fact of *there being no bottom;* the steep side of the gorge terminating in a stream of water into which the animal was softly but rather precipitately landed. The soup diet is beginning to show itself in the aspect of the men, who are becoming lean and emaciated; while the horses are becoming weak and jaded from starvation and overwork. Even "water had become scarce in these horrible mountain deserts and it was with as much joy and rejoicing among the corps as happens among passengers at sea, who have experienced a dangerous and protracted voyage, when they first discover land on the long looked for coast," that they beheld, at last, a level plain in the distance.

Even horse flesh became scarce and so dry from want of nutrition, that it was little better than sole leather, the carcass of a wolf was a rare delicacy and the incident of one of their hunters procuring a supply of berries, roots and fish from the Flatheads as matter to be thankful for. As they emerged from the mountains, their route became gradually more comfortable. In a small valley, they found a village of Flathead Indians, who supplied them with provisions, consisting of fish, roots and bread, manufactured by them in a way peculiar to themselves, from a sweet root, growing in great abundance on the plains, and bearing in June a flower of a pale blue color, the root resembling the onion in appearance, which they call "comas." This bread was found not only nourishing but quite palatable, tasting like that made of pumpkins. The Indians treated them kindly, and furnished them with a good supply of edibles in exchange for small articles of merchandise, which they had brought along. From this point they travelled by moderate stages, having many of them fallen sick from bad and insufficient diet, and it may refresh the memories of our readers to be informed that Captain Clarke in this emergency with true Sangrado policy "gave all the sick a dose of Dr. Rush's pills to see what effect that would have." The experiment appears to have been successful, Dr. Rush's pills did their duty, and the men began rapidly to improve in bodily health and spirits. Game continued scarce on the Columbia, the Indians of this country having to cross over on to the Missouri to procure their supplies of meat which they annually did in large parties in pursuit of the Buffalo. Another kind of native sheep is spoken of as living in these vallies, distinct from the mountain Ram of the preceding pages in being smaller and covered with wool four inches long, fine, white and soft; instead of the hairy covering peculiar to the latter animal. The want of an interpreter to enable them to communicate with the Flatheads proved a serious inconvenience, but they managed by signs to ascertain that they were then at war with a neighboring tribe, who had slain several of their people; and also, that they had had communication with white men at the mouth of the river, upon which they were then travelling.

It is amusing to notice the indifference with which, our author, by this time has learned to speak of dining on horse flesh. The hunters, came across a horse, shot him and after dressing, hung him up as if he had been a Buffalo or a bullock, and the party partook of his flesh, with even more gusto perhaps, than do the Parisian gourmands—to a certainty, these latter, have not such an excuse for an appetite. Game was utterly out of the question. The water was also warm and soft and sickened them. The very timber on the mountains was dead and fallen and starvation seemed to be the genius, of the place. The only redeeming feature mentioned is a kind of rock, suitable for millstones. They here, supped upon the last of their horse, and by way of desert, finished on a wolf they had killed; and which, Mr. Gass, calls very strong and substantial diet. After this, they came into a section where berries and Indian bread abounded, but the change of diet made the men sick and they were forced to lie up and recruit.

By the 1st. October, the men had all sufficiently recovered to be able to work, and the navigation had so much improved, that it was deemed advisable to prepare canoes, and continue the journey by water. The labor of hewing out canoes was thought too arduous considering the weakened condition of the men and accordingly they were hollowed out by burning in the Indian fashion, which process consumed some days; and on the 8th. they were again prepared to continue their voyage. Along the river they discovered numerous lodges of Indians, who were uniformly peaceably disposed, and two chiefs who volunteered to accompany them, gave them the cheering assurance that ere long they should meet white people, and as evidence, of the fact they displayed beads and other trinkets of white manufacture.

At the mouth of the Koos-coos-kie, a large river coming in from the east, their Snake guide deserted them, frightened by the difficulty of the navigation. The principal portion of the men of this band of Flatheads having been on a war party, returned about this time, and came to the camp of the whites, but owing to the absence of an interpreter they were unable to give any information. They were, however, very peaceably disposed, and having received various present, remained loitering about the camp. Provisions of a suitable kind continued very scarce and more horses were slaughtered, though the natives supplied them with an abundance of their kind of provision. What horses were left, were got together and branded in the forlorn hope that they would be forthcoming on their return; and leaving them in charge of an old chief of the Flatheads, they embarked their canoes on the bosom of the Columbia.

The operation of flatening the heads of the young Indians, is given as follows, by Mr. Gass:

"This singular and deforming operation performed in infancy in the following manner: A piece of board is placed against the back of the head, extending from the shoulders some distance above it; another shorter piece extends from the eye-brows to the top of the first, and they are then bound together with thongs or cords, made of skins, so as to press back the forehead, make the head rise at the top, and force it out above the ears."

The country on this portion of the Columbia was high, dry prairie, with scarcely timber enough to furnish firewood with which to cook, but of average fertility; the hills rocky, but not very high, and the stream rapid and clear, the bottom covered with stones of an uniform round shape. The prevailing food was now fish and dog-meat; owing to scarcity of salt, the former was insipid, as well as unwholesome, and the men much preferred the latter, which, says our author, "when well cooked, tastes very well." Large numbers of dogs, as well as horses, are kept by the Indians, and these animals are slaughtered and eaten with as much nonchalance as our butchers would kill a bullock car a sheep, and our voyagers came gradually to prefer dog-meat, to almost any other kind of provisions the country afforded.

As they passed down the river, the navigation rapidly improved, and were it not for the rapids, the Columbia would be a beautiful river, while the country, except for its uniformity, had quite an attractive appearance. Says Mr. Gass: "This river in general is very handsome, except at the rapids, where it is risking both life and property to pass; and even these rapids, when the bare view or prospect is considered distinct from the advantages of navigation, may add to its beauty, by interspersing variety and scenes of romantic granduer, where there is so much uniformity in the appearance of the country."

At the mouth of the North West fork of the Columbia, called by our party the Great Columbia, the country all around is level, rich and beautiful, but without timber. The width of the river at this point, is 860 yards, while the lesser fork, called now Lewis' River, is in width, 475 yards. After the confluence, the Columbia becomes a majestic stream, its course interrupted by frequent rapids, but navigable otherwise for large vessels. Its waters swarm with salmon and other fish which furnished subsistence to numerous Indians, who inhabited its banks, but who, as represented by Mr. Gass, were nearly naked, and in a miserable condition. In regard to the natives, says the Journal, "there are three, or part of three, different nations here. They are almost without clothing, having no covering of any account, except some deer skin robes and a few leggins of the same materials. The women have scarce sufficient to cover their nakedness. They appear to be very shy and distant. On the 19th, a number of the natives came to our camp, and our commanding officers presented one of them with a medal and other small articles. We also passed a great many Indian camps; and halting opposite a large one, about thirty-six canoe loads of them came over to see us; some of them remaining all night; but we could not have much conversation with them, as we did not understand their language. They are clothed much in the same manner with those at the forks above. The custom prevails among these Indians of burying all of the property of the deceased, with the body. Amongst these savages when any one of them dies, his baskets, bags, clothing, horses and other property are all interred: even his canoe is split into pieces and set up around his grave. Next day, we came to an Indian camp on the point of a large island, where stopped and got some fish and other provisions. We here saw some articles which showed that white people had been here or not far distant during the summer. They have a hempen scine and some ash paddles which they did not make themselves. At 1 o'clock, we proceeded on again, went forty-two miles, and encamped without any of the natives being along, which is unusual on this river. We could not get a single stick of wood to cook with and had only a few small green willows.—We continued our voyage, and at an early hour came to the lodges of some of the natives. Here we got some bread, made of a small white root, which grows in this part of the country. We saw among them some small robes made of the skins of gray squirrels, some raccoon skins, and acorns, which are signs of a timbered country not far distant. Having proceeded on again, we passed several more lodges of Indians; and through two very rocky rapid parts of the river with great difficulty. The next morning was fine, and we saw a great number of ducks, geese and gulls. At 10 o'clock we came to a large island, where the river has cut its way through the point of a high hill. Opposite to this island a large river comes in on the south side, called by the natives Sho-sho-ne or Snake-Indian river; and which has large rapids close to its mouth. This, or the Ki-moo-ce-neim, is the same river, whose lead waters we saw at the Snake nation. The natives are very numerous on the islands, and all along the river. Their lodges are of bulrushes and flags, made into a kind of mats, and formed into a hut or lodge."

On Wednesday, the 23d October, 1805, they reached the rapids or great falls of the Columbia, the first pitch of which is 20 feet perpendicular, being thirty-seven feet in a distance of twelve hundred. The water sometime reaches to a height of forty-eight feet, at which times, the falls become only a rapids and can be safely passed over with boats. At ordinary times, the channel is only seventy feet wide for some three miles; and the immense mass of water being thus confined, rushes with almost lighting velocity. About the great pitch, the appearance of the place is said to be terrific. Tremendous rocks threaten to topple over with the trembling of the earth, and the mighty volume of water pouring over into so contracted a channel lashes itself into foam and fury. The waters seem in torment and the beholder invariably feels creeping upon him a sensation of awe and even of fear, of so indefinable a nature, that he involuntarily shrinks from the contemplation. For a considerable distance from this point continuous navigation was rendered impossible by similar obstructions; and the party was forced to carry their canoes and loading by land--sometimes for miles at a stretch, and thus slowly and laboriously, they pursued their difficult way over this portion of the river. At length, the current of the river became more uniform and they were enabled to make better headway, holding occasional conferences with the natives, from whom they learned that a conspiracy was being formed among the Indians farther down, to waylay and exterminate them; and subsisting on dog, dried fish, and such other delicacies as they were able to procure from the Indians. Occasionally, a deer was killed and brought in by the hunters, while water fowl was quite abundant. Obviously, their situation was improving, as they descended from the inhospitable mountain country into the fertile bottom lands of the Columbia. Monday, November 4th, opened up fine, clear and frosty; and the portion of the river they were in, excited their admiration by its beauty; but more cheering even than the contrast of such a river, with that over which they had made such toilsome progress, was the fact revealed to their vision, that the river rose and fell with the tide, and the information, conveyed by signs by the Indians, that in two more days they would see ships with white men in them. As evidence of their veracity, they displayed quantities of new cloth, and of trinkets they had obtained from the ships, and the despairing mariners of Christopher Columbus, never viewed with more solicitous curiosity or more heartfelt satisfaction the floating evidences of the land they were seeking beyond the western waters, than did our adventurers these signs and symbols of a civilization to which they had been for so many weary months such total strangers. At length, on the 16th. November, 1805, they saw for the first time the waters of the Pacific. For some days there had been almost a constant storm, and the bay at the mouth of the river was turbulent, and rough; so that their first impressions of the great western ocean was anything but favorable as to its pacific character. All the reflections, our journalist, who is as sententious as Tacitus, on such subjects, has to make on an event, which might well be considered an epoch in an ordinary lifetime, and with a more ambitious author might have excused some self glorification, are embodied in the following short quotation, the brevity of which is only surpassed by its exceeding modesty.—"We are now at the end of our voyage, which has been completely accomplished according to the intention of the expedition, the object of which was, to discover a passage by the way of the Missouri and Columbia rivers to the Pacific ocean; notwithstanding the difficulties, privations and dangers which we had to encounter, endure and surmount."

There appears to have been very little romance or sentiment about any of the party, all such unsubstantial ideas having been starved out by hard, practical experience; as the next intimation we have of their proceedings, is, that five of them went out to hunt and returned with so many deer, ducks and geese; while the balance quietly sat down to wait for Captain Lewis, who with some men had gone in quest of the white people of whom the Indians had informed them by signs. The broad Pacific rolled before them in its turbulent majesty: at their backs, frowned the mountains whose fastnesses they had dared and whose secrets they had learned: while at their feet, lay a fertile land of boundless extent, watered by mighty rivers and in a genial climate but in unclaimed and savage wildness; but they threw neither fetters in the sea or planted stakes upon the land. There was no planting of crosses, no advancing of banners, no ceremonies to commemorate the occasion, such as other explorers had deemed necessary when a country was to be wrested by the grace of God from its natural owners, and transfered by a flourish of paper, burning of gunpowder and sacrilegious calling upon

Deity, to his catholic or his protestant majesty: but in a plain matter of fact way they went about their business, seemingly unconscious that they were the pioneers in the greatest Exodus that has ever happened since Jehovah himself, led his chosen people from the land of their bondage into a country flowing with milk and honey.

    Like the Israelites of old, full forty years elapsed before the fruition of hope; and all the wanderers, save one, were in their graves, before the land they discovered became in reality the land of promise. Mr. Gass, alone survives, the sole living testimony to a modern miracle, almost rivalling in its wonderful sequences the journey through the Red sea and over the desert wilderness led by the prophet of God. Forty years after him, a living stream of adventurous men began to pour into the vallies of California and Oregon; they swarmed over the sterile plains and scaled the mountain passes, and their sails whitened the bays and harbors of the coasts. The wild Indian looked on amazed, and the haunts of the buffalo and grizzly, echoed with the shouts of teamsters and the creaking of loaded wains, as company after company and drove after drove pursued their wearisome way, impelled as it were, by the hand of Providence, to settle and thus subdue this modern Canaan. In ten years time, cities, villages and hamlets sprang up; the Golden State was organized, and peopled with an enterprising, intelligent population and added to the great confederacy, whose domain was thus made continental. California, the result of this grand irruption, although but an infant in years, has already outstripped some of her older sisters in all the attributes of greatness--numbers, wealth and intelligence; and other embryo states are knocking for admission in the mountains of Oregon. Singularly enough, the pillar of fire by night and the cloud by day, in this modern Exodus, was gold,--gold in the dreams and gold in the daylight visions of the thousands of every name and clime, who now people the Golden State, or whiten with their bones the same plains and sierras skirted and traversed by our adventurous party. It does seem indeed as though the hand of Providence were in it. For thousands of years the yellow metal had reposed, waiting in the sands, the time when all things conspiring, it should be disclosed to tempt the cupidity of man, and accomplish in the settlement of the country the beneficent designs of the Creator. The world was all at peace, and unexampled prosperity hovered over all the nations of the earth. Commercial enterprise was in its amplest development, and the spirit of speculation was rife in every land. People were just ready for such a discovery of gold. The disclosure broke upon them like the news of a panic, all listened, all believed--few reflected--and many ventured.—Scarcely a nation on the earth, but was soon represented in California. Natives of the Celestial Empire landed from their junks; barbarous islanders from the Pacific; Africans, Asiatics, Europeans, and Americans, all concentrated upon her shores in the rush after the golden prize. The sequel has demonstrated, it is true, that all is not gold that glitters, but bas proven in the far searching providence of God a world-wide blessing. Of all this grand development, our party had no idea, and probably had a prophet risen from his grave to reveal the future, they would have treated him with incredulous scorn. Such is short sighted man, with all his knowledge, all his sagacity, all his courage and his pride.

    The whites referred to by the Indians had departed shortly before they arrived, leaving them and the Indians sole monarchs of the domain. Capt. Lewis discovered where they had encamped, but our author gives us no information as to the nation or character of the ships, referred to, though more than probable they were Yankee whalers, who had put in here for a little dicker with the Indians during the trading season.

Having reached the mouth of the Columbia, after traversing over four thousand miles, of unexplored wilderness, and expending eighteen consecutive month; in the operation, it became advisable to take measures for spending the winter season as comfortably as possible, before commencing their return in the spring. November was far advanced and the increasing inclemency of the weather, warned them to be on the alert. During a month spent at the month of the Columbia, reconnoitering the country, they experienced only three fair days and it was not until the 5th. of December, 1805, they were able to pitch upon a spot that suited their purpose; and they immediately proceeded to move their effects to the place, a distance of some fifteen miles up a small branch coming into the bay, where they found game in considerable abundance, and the facilities for making salt, of which they stool greatly in need. Elk were seen in large numbers, and quite a number were killed by the hunters of the party. By Christmas day, their winter quarters were completed, being made of puncheons and logs comfortably daubed with mud, and the men left their hunting camp and moved into them. On Christmas morning all the men paraded, and firing a round of small arms, wished the commanding officers a merry Christmas. This appears to have been a kind of superogatory wish, as our author intimates that the article with which to *make merry* the heart of man had long since vanished, but the officers in the true spirit of courtesy accepted the will for the deed, and in lieu of grog, collected what tobacco was left, and divided it among those who used the weed, by way of Christmas gift; while those who did not, had to solace themselves with *cotton handkerchiefs*. The party were now all in excellent health; with plenty of meat, and generally well provided for, except that they had no salt, owing to the want of which, a great deal of their meat was, spoiled. Although in so northern a latitude and at so late a season, the weather still continued warm enough to allow ticks, flies, and ocher insects to exist in annoying abundance, and it was almost unintermittingly rainy. January and February, wore away, with nothing remarkable to disturb the monotony of killing elk, making salt and preserving the meat, unless the incident of a dead whale 105 feet in length, washed upon the beach, be considered of sufficient importance to bear narration. This state of affairs, continued until about the 1st of March, when it was determined that they should set out on their return to the states. It may be supposed that this determination was viewed with an unanimous approval and that visions of welcome home by friends, kindred and sweethearts, and of that honorable estimation for daring and perseverence, so dear to ambitious and adventurous characters, and that prompts men to seek the bubble even at the cannon's mouth, warmed the hearts of our travellers into something of a glow as they again took the trail for the far country away to the east of the mountain ranges on the farther slope of the continent. It is natural to imagine that men under such circumstances would indulge in some such visions, and we will credit them with enough of common human sympathy to suppose such a case, but strict historic truth warrants no such a pleasant fiction from anything found in the pages of our author. Long acquaintance with Indian habits had apparently induced a stoic pride, which forbade manifestation of feeling by words; and the page is as destitute of reflection, gratulation or of any exhibition of human feeling, in any shape, as the rocky slopes of the savage mountains were of cheering verdure. We have the naked record, that without any particular stir, they left their encampment about the 1st. of March 1806, and journeyed by slow and irregular stages up the Columbia river. The journey up this river is meagre of incident—being merely a repetition of what occurred during the descent. They subsisted on game, which they found in abundance; and on the dog meat, with which the Indians abundantly supplied them. Their long deprivation from the luxuries of civilized life, had had its effect upon their physical as well as their mental and social nature, and the food which would revolt the stomach of the pampered dwellers in our land of ease and plenty had become to them not only nourishing but savory. It is curious to observe the effect of circumstances upon the tastes and characters of men, and the result of the observation will be that man of all living animals possesses not only the most pliable of constitutions, enabling him to surmount all hardships and privations; but that his very nature can be so changed and made to conform to the features of the surrounding circumstances, that he may become in time radically distinct from his blood kindred. Thus it is, that the various races of men have increased upon the earth, which, philosophers for the lack of a better phrase have denominated varieties;

and hence, in the various Indian tribes of the American continent, amounting to some hundred, no two are so similar but that they may be easily distinguished by physical marks, which every Indian could recognise. In fact the trappers and hunters of the western prairies become themselves a species of red-men, not farther removed in appearance, habits and speech from the true Indian type, than from the white stock from which they spring. Time and circumstances we have reason to believe, would make them and their descendants as much Indian as the Camanches or Flat-heads among whom they exist.

About the Falls of the Columbia, the crossing of which was effected without any particular incident, Mr. Gass speaks of observing on the plains a "species of clover as large as any he had seen in the States, and bearing a large red blossom." The leaves, he says, were not quite so large as those of the red clover of the States, but more abundant, being from six to eight on a branch, whereas the latter has but three. He speaks in high terms of the appearance of the country in this vicinity, under the genial influence of the spring's alternate sun and showers. In the distance to the southwest, was to be seen a range of snow clad mountains, glittering in the sunlight, a sad reminder of the difficulties they were yet to encounter, while at their feet was a soft emerald sward, bedecked with gay flowers, and gathering additional beauty from the contrast. They halted at this pleasant spot for some little space, and were entertained with a grand dance by the Indians, who flocked to see them from all quarters. After procuring a supply of dogs, with comas roots and Shap-pa-leel for provisions, on the 1st of May they resumed their line of march toward the east. Meeting with considerable difficulty in procuring a proper supply of eatables, they travelled up the Columbia, passing the junction of the Kooscoos-kie, on the 6th of May, 1806, and recovering on their route, several of the horses, which they had left in the care of the old Indian, on their way over; and which were punctually returned, and acting as physicians for the Indians, who had as high an estimate of the white man's skill in medicine, as modern pill venders have of the efficacy of Indian remedies—and generally rendering themselves agreeable to the natives—which was rewarded lay many kind offices on the part of these latter. Mr. Gass' says that "all the Indians from the Rocky Mountains to the Falls of the Columbia, are an honest, ingenuous and well disposed people; but from the Falls to the sea coast, and along it, they are a rascally thieving set." Chastity in his opinion, seems to have been considered a virtue among none of the tribes.—As they ascended the slope of the mountains they experienced considerable difficulty from the snow which they found several inches deep and still occasionally falling and provisions very scarce, so that they were frequently obliged to kill and eat their beasts of burden as well as dog-meat and roots. They therefore acting under the advice of the Indians, concluded to delay a short while, during which time the snow might become sufficiently melted to allow of crossing the Mountains. The time here was improved by the hunters in procuring meat; and by the officers in the practice of the healing art among the Indians, numbers of whom were brought by their friends for the benefit of their services. On the 15th. they left this place, called the "Commas-flat," the first place where they had found any of the natives, the fall before, after crossing the mountains; and which, is represented to contain about 2000 acres of land, covered at that time with strawberries in blossom and surrounded with excellent pine timber of various kinds. They had now sixty-six horses all in good order and were again tolerably well stocked with provisions. They found the snow in the mountains varying from five to fifteen feet deep, entirely obliterating any track and rendering it dangerous as well as impracticable to proceed without a guide. In this emergency, they were forced to turn back, disappointed and melancholy. Notwithstanding the snow in such troublesome proximity, the mosquitos and gnats were extremely annoying, compelling them to build small fires to protect the horses from their attacks.

At length, on the 1st. July, 1806, they had passed the more difficult portion of their route, crossing the mountains, and halted to rest at the mouth of Clarke's river. The party, was here separated; a part going up this river, with Captain Clarke; our hero under the command of Captain Lewis, with several others having to go straight across to the Falls of the Missouri, where they had left some canoes. On the 3rd. July, they started—Captain Clarke up the river and Lewis and his party, with the accompanying natives, down.—They here dismissed their guides with many presents, and Mr. Gass, again highly compliments these Indians, as "hospitable, obliging, and good hearted sons of the West."

After wandering around through the broken country lying between the waters of the Columbia and the Missouri, our explorers on the 7th, came upon the dividing ridge which finally separated them; and starting from a mountain spring, they followed its course, day after day, until on the 11th. they struck the main river near the scene of their encampment the winter before. A few days were spent at this point in looking up their baggage and boats concealed previous to crossing toward the west; and Buffalo and other game being very abundant it was considered advisable for the larger portion of the party to remain and lay in a stock of provisions; and make such arrangements as might be advisable previous to attempting the descent of the Missouri; while Captain Lewis, with three hunters would ascend and explore the section of country, lying on Maria's river.

His instructions were to await his return at the mouth of Maria's river, until the 1st of September, at which time should he not arrive, they were to proceed on to join Capt. Clarke at the month of Yellow Stone, and continue thence homeward; but he informed them, that if "life and health be spared, he would meet them at the mouth of Maria's River on the 5th of August."—The Captain departed on his uncertain mission, and our hero and the larger portion of the party remained in camp, occupied in hunting and repairs. The bears were bad and one occasion, Capt. Lewis came into such close quarters with one, that he broke his gun over bruin's head, and while the animal was recovering from the shock, found opportunity to climb a tree, where the animal besieged him for three mortal hours. However, bruin's patience at length gave way, and the Captain, duly thankful for his safe deliverance, descended and caught his horse, which by the way had taken fright and thrown him almost into the teeth of the bear, about two miles off, and made the best of his way to camp.

Sunday the 27th, found the party duly provided with provisions and conveyances at the month of Maria's river, and quite unexpectedly they met Capt. Lewis, with his three hunters, who had had a skirmish with a party of Grosventre, or big-belly Indians. They had encountered the party, who appeared very friendly, exchanged presents and passed the night with them peaceably enough; but the next morning, they suddenly snatched up the guns of three of the whites, and made off with them, the whites followed, and one of the Indians was killed by a stab with a knife, and another mortally wounded by a shot, the whites escaping unharmed and recovering all their arms, besides coming into possession of a number of horse, which the Indians abandoned in their flight. The experience, however, satisfied the party as to the exploration of Maria's river, and making all haste toward the mouth, they reached it at about the same time the main party arrived, according to appointment.

On the 29th of July, having perfected all their arrangements, they turned their horses loose on the plains to take care of themselves, and embarked in their canoes to descend the Missouri. The river being high and rapid, their descent was rapid, and comparatively without adventure, beyond the occasional killing of a bear, and the ordinary incidents of hunting experience.

On the 7th, arriving at the mouth of Yellowstone, the appointed rendezvous with Capt. Clarke, they discovered that he had gone some time before, and left no trace, except some few words written in the sand, stating that he had gone a few miles farther down. They followed, passed several of his camps in succession, and on the 12th, overtook him and his party, all in good health and spirits, and piously ejaculates our hero,—"thank God, we are all together again." Their journey was now drawing to a close, after having endured hardships and uncertainties of an expedition unexampled for the length of time occupied, the territory traversed and the successful prosecution of the same, they had at last all got again together, and were speeding with light hearts and glad anticipations toward their own, yet distant homes. No further difficulties need be anticipated, and a very few days would again enable them to see the welcome faces of white men and resume the almost forgotten customs of civilized life. Already the vanguards of the white man were around them, and daily they passed or overtook trappers who were following their vocation among the Indians of the Missouri, and from them they received their first news, albeit, a twelvemonth old, of the occurrences, the changes and revolutions that had occurred during their protracted exile.

    Among all their privations, none seem to have affected them worse than that of tobacco, and accordingly the opportunity to exchange a boat load of corn with a St. Louis trader for a supply of the comforting weed, was a source of exceeding joy, and thought worthy of commemoration in Mr. Gass' Journal. Their first call was for tobacco. Say what we will, about the folly or the evils of the use of this article, there is certainly a charm about it, which to properly appreciate, one must submit to a long deprivation. Alike to the sailor, the soldier, the traveller, the trapper, as well as to the man of more steady habits in settled life; it is a comfort in fatigue, a stand-by in distress; and a promoter of good will, a peace maker in argument, and a friend in all emergencies, especially those requiring a quick intellect, a cool head and a resolute will. From another St. Louis trader they procured a supply of

Monongehela whiskey, the first spiritous liquor they had tasted since the 4th of July 1805, just previous to undertaking the eastern ascent of the Rocky Mountains.

From this time until the 23d of September, when they arrived safely at St. Louis, nothing particular occurred, unless the meeting of a trading company commissioned by the government, to make enquiry concerning their whereabouts, be considered noteworty. Their long absence had somewhat disquieted their friends at home; and the government were about taking measures to enquire as to their welfare, when very opportunely, they met the messengers, and in the most satisfactory manner relieved them of all disquietude. On arriving at St. Louis, then, the rendezvous of the Indians and of the bronzed and bearded trappers of the northwest, for the purpose of trade and procuring supplies, they were of course the lions of the day. Their appearance, tanned and grizzled; hair and beards uncut, unkempt; attired in leathern suits or garments of skin, and adorned with Indian ornaments, was sufficiently outlandish to excite remark even in that theatre of outlandish costume; but the intelligent account they could give of the country they had traversed, the superstitions and exagerations they dispelled in regard to the customs and numbers of the Indian tribes, the specimens they brought home with them of the animal and vegetable products of the country, gave them an importance, leaving out of the question their official character, that secured them the highest respect. The commanding officers had kept Journals of the details of the expedition, which were published at great expense by the government, and copies presented to foreign governments as great accessions to the knowledge of mankind; while the more intelligent of the men were also enjoined to keep a record of events, so that in case of accident the chances of an authentic account of the expedition, might be increased. Acting upon this direction, Mr. Gass, kept a diary of events, which was afterwards arranged for the press by a Mr. David McKeehan, and published at Pittsburgh in 1807. From this work, which as the publisher informs us, was but very slightly altered, either in verbiage or arrangement from the original, we have drawn largely in the preceding pages, culling the leanding facts, condensing the material portions, and adding incidents and reflections on subsequent occurrences, to suit the taste of the modern and desultory reader of such travels. The original, gives evidence of close observation and of much shrewdness of reasoning. It is, we believe, strictly and conscientiously accurate, for contrary to the received aphorism in regard to travellers tales, we have never perused a work so devoid of the imaginative or where was manifested so little desire to garnish plain prose with poetic tinsel. All is plain unpretending matter of fact, just such notings as a mathematician might make in a scientific traverse of the land. We see the adventurers just as they were, and with rare modesty, the author, although we have authority for saying that he was one of the most useful, efficient and intelligent men of the party, is kept strictly in the back ground, or if mentioned at all, it is only incidentally and in connection with some special party of which he was a member. This is always to be considered a characteristic of true merit, and has usually attached to those men who have most distinguished themselves for sterling qualities. There is a foppishness about some great men even in the article of modesty, which shines through its flimsy disguise, in spite of all their efforts; but with others, there is a real unaffected naturalness of demeanor, that we instinctively recognise and appreciate. Caesar, in his commentaries is a sample of the former; while Washington, in his whole career is a specimen of the latter. Caesar, by an affected translation of personalty, transfered himself into the third person, and told most marvellous stories in a plain way, of which he is always the hero, in *vini, vidi, vici* style; while Washington left his history to posterity and was scrupulously exact in all his official narration, scarcely seeming to regard himself as an agent, but still leaving impressed upon the mind of the reader, the conviction that he is the moving spirit. We do not wish to institute a comparison between our hero, and those illustrious characters; but his character in its indomitable will, great self reliance, calm courage and unaffected modesty, was more of the American than of the Roman mould. All these characteristics are strikingly apparent in his career, as set forth in the unpretending pages of his Journal; and in his subsequent life, he followed the bent of the same inclinations.

Remaining at St. Louis but a few days to receive and enjoy their honors, and the hospitality of the citizens, the party proceeded east to make report and obtain their discharge. Mr. Gass, travelled by land to Vincennes, Indiana, and awaited there the arrival of Messrs. Lewis and Clarke, who followed with a deputation of Indians from the plains, among them a chief named Big White, whom Mr. Gass calls the best looking Indian he ever saw, which, they were conducting to Washington City, for the purpose of demonstrating to them by observation the overwhelming power of the United States and the uselessness of hostility on the part of the Indian tribes in case of any dissatisfaction with the government on their part. The lesson was designed to teach them prudence, and as the wild sons of the prairie, journeying through the land of the palefaces, dwelt upon their cities and villages, and noted the number of the whites, the great resources of the nation for peace or war, and looked with admiring wonder upon the long rows of stately houses, the heaps of glittering goods, the public edifices, fortifications and shipping, so striking to their unaccustomed eyes, the conviction of the white man's power forced itself upon their minds, mingled with prophetic forewarnings of the red man's fate. They looked on with a sullen and stoic indifference, but not a sight or a motion escaped their gaze. Their observations, doubtless, have had their effect in determining the conduct of their wild brethren of the West.

The commanding officers, having changed their route of travel, Mr. Gass, with a couple of companions, proceeded to join them at Louisville, Kentucky. Among the Kentuckians, they were received with the highest honor, citizens of all classes exerting themselves to make their sojourn among them as pleasant as possible.—Among the entertainments, here, in their honor, was a grand fancy ball, which they all attended; their Indian companions tricked out in all their savage finery, with necklaces of white bears claws, brilliant brass medals and gorgeous plumage and painting. The curiosity of the whites was excited to the highest pitch, not only to see the members of the party but to inspect the curiosities they carried with them as trophies. Through the whole route they were the objects of marked attention; and as they came into the more settled portions of the states, their progress almost

BIG WHITE — BALL COSTUME. — [Page 108.]

resembled a civic triumph. It may be said, that Lewis & Clark united the Atlantic and the Pacific, as Cyrus W. Field, did continents, in the bonds of science; and the latter achievement was not accompanied with more laudations than the former. They at last reached the Federal City, and after paying their respects to President Jefferson, making their report to the proper officials, delivering over their specimens and curiosities they were discharged with a vote of thanks and a worthy acknowledgement of their meritorious services.

Mr. Gass received his pay in gold, with the promise of future consideration at the hand's of the country, and set about enjoying it at his leisure; and during the next few months of his career, we have no information of his proceedings except that he returned to his friends in the vicinity of Wellsburg, and spent a few months in comparative inactivity.

Of the subsequent history of his commanders, Captains Lewis and Clarke, we have but a meagre detail, and still less of that of his companions in the ranks. The officers were both men of more than ordinary ability and qualifications, and afterwards attained to very respectable public station. Lewis was appointed very shortly after his return in 1806, Governor of Louisiana territory, as some acknowledgement of his merit, and compensation for his services. In this capacity he acted for sometime, but unfortunately a misunderstanding arose between him and the government in regard to the settlement of his public accounts. He was the very soul of honor and of unimpeachable integrity, and the implied imputation, dwelt too heavily upon his proud and sensitive spirit. He started to Washington City for an explanation, but never reached his destination. In company with another man he travelled the old route followed by the boatmen at that day, through the Indian country; and having reached a small cabin occupied by a man named Grinders, as a kind of tavern for travellers, just within the Chickasaw nation, near the Tennessee line, and between twenty-five and thirty miles of Nashville, his man left him to go in search of a horse that had strayed. During his absence after the horse, Lewis shot himself twice with a pistol, and this failing to effect his purpose, he killed himself by cutting his throat with a knife. No one saw him commit the act, but some of the family afterwards reported that they had observed indications that his mind was affected on the morning of his death. His body was buried at the corner of the cabin, and for a long time after, the spot was remembered by the adventurous traders who passed that way, between New Orleans and the upper country.

Thus was ushered into eternity a brave and chivalrous spirit, goaded to desperation by the chafing of wounded honor. His untimely death was universally regretted. Who can describe the poignant anguish that could have impelled such a man into the commission of such an act—an act from which the mind recoils with instinctive horror. Peace be to his memory. The great Arbiter of all be the judge of his motives, as He alone must be the dispenser of his deserts in the land of the dread unknown, into which, all unannounced, his own rash hand ushered his living soul. It is enough for the historian to say that he died with the cloud upon his memory; and while he records his fate with a careful pen, he would ask of the world its most charitable judgment. The charges against him were hushed, communities and states vied to do him honor, and the Legislature of Tennessee, his adopted State, to manifest an appreciation of what was high and noble in his character and services, ordered a monument to be erected to his memory at the State's expense.

His associate Clarke, received the title of General, and in 1813, just at the commencement of the war, received an appointment as Governor of Missouri territory and Superintendent of Indian affairs, an office of great responsibility and importance in view of the impending war, and of the evinced determination of the British Government to array against us the horrors of Indian warfare. His selection for such a post is an indubitable proof of his standing. He continued to hold these offices with acceptability throughout the war, and until the admission of Missouri as a State in 1820.—In 1822, he was again appointed Superintendant of Indian affairs, and held the office for many years afterwards. In the mean time he had married, and had his residence at St. Louis, where he raised a family and died in 1838. His remains were followed to the grave by an immense concourse of citizens, strangers and Indians from the plains and mountains, and is said to have been the largest funeral ever witnessed in St. Louis.

The results of Lewis and Clarke's expedition have become matter of history: their contributions to science, having now been merged in the great mass of the intelligence of the country. They all have gone to their last account except the subject of our memoir—who yet lingers, tough and gnarled by time, on the verge of that great wilderness he must soon in the order of things be called to explore, in the world to come. If the foregoing pages shall serve to stimulate some one, to emulate his patriotism or excite one generous glow of admiration of his unselfish character, in the bosom of a single reader; of his untiring zeal in the discharge of duty, his modest deportment under all circumstances, or of his indomitable will, the object of the writer will have been in that much attained. We are now drawing to the close of the most important era in his life, and after a few desultory remarks upon the modern aspect and history of the scene of his travels, we shall proceed to narrate his subsequent career.

The route traversed by them, has never been of much practical advantage as a means of communication between the Atlantic and Pacific, being too far to the north, and much more available passes through the Mountains have since been discovered; but their success, demonstrated the practicability of a passage and served to stimulate subsequent explorers. The Rocky Mountains since their time have lost much of their terror. The route travelled by the emigrants to California and Oregon, by way of the Platte and Kansas rivers, Salt Lake city, Bridger's Fort and the South Pass on to the waters of the Sacramento and the Columbia, is of very gentle ascent; and presents no greater difficulty than do some of the routes over the Allegheny Mountains, that are now traversed by roads and railways. The South Pass, so much used by these emigrants is not far from the crossing place of Lewis and Clarke, they having just missed it by keeping too far to the north. Near it, is Fremonts Peak, 1300 feet in height. The Pass actually discovered by them is barely practicable and never used. It was not until the discovery of gold in California that attention was directed, in earnest, toward this portion of the world; but in a very short time after that event, the whole region was thoroughly explored.—The voyage by sea was both costly and dangerous and it became necessary to find some available route by land. Private enterprise and thirst for sudden wealth soon effected it; the wave of emigration sweeping up to the base of the Rocky Mountains soon found its level and following up the vallies and gorges of the mountains trickled through their fastnesses in many a winding stream, until gradually it settled into the well defined channel that is now almost as well known and as well worn as is any thoroughfare in the states.

A new impetus was given to the spirit of discovery in these regions on the development of the magnificent scheme of the Pacific Rail-road. The merit of originating this idea, is generally attributed to Mr. Whitney, of New York, who in 1844, first definitely broached it before congress. His idea was to connect the valley of the Mississippi with the Sacramento, the Columbia or the Colorado, by means of a railroad according as the most available route might be found; the expense of making the road to be defrayed by appropriating to contractors alternate sections of the public lands on either side of the road. The plausibility of such a scheme may be seen at a glance, but it was a gigantic undertaking; and its possibility even, had not yet been reliably demonstrated. Hon. Thos. H. Benton, early became a patron of the project, and gradually it forced itself upon the attention of Congress and the public.—Whitney, himself, was an enthusiast in the cause, and just at that period, railroad speculation was at its height throughout all the States of the Union. Able and voluminous reports and speeches were made on the subject of the feasibility of the Pacific Railroad--it forced itself into the messages of the Presidents—and into the arena of politics, and apparently the dream of its projector was about to be realized. Foreign capitalists embraced the scheme, and promised their assistance to effect its consumation. The brilliant services and favorable reports of Fremont, who was engaged during 1845 to 1850 in a semi official capacity in exploring the country, contributed to heighten the feeling in favor of the road, and demonstrate its practicability. A damper, however, was put upon his representations in the winter of 1848-9. Allured by the tempting openings as descried from a distance in the Sierra Nevadas, he was with his party caught in a snow storm in the mountains, and barely escaped with his own life, leaving some of his comrades and all his animals and effects victims to the frost and snow. The celebrated Christopher Carson, was a companion and guide of Fremont's during these explorations and by his indomitable energy and great sagacity rendered himself equally conspicuous with his superior in command. This misfortune, which happened to the south-west of the great Salt Lake, and near the line of travel to San Francisco, only seemed, however, to attract attention to the country. The Mormons, driven from their homes in Illinois and Missouri about this time, were founding their State of Deseret; with Salt Lake City for a capital and a bee-hive for their coat of arms. In all quarters of the States and in the old world, they listened to the voice of their prophet, and pouring into their new found city of rest, hoped to build up there, a peculiar nation sacred from gentile intrusion. The Mormon settlement at Salt Lake City filled up rapidly with the deluded followers of Brigham Young— Governor, by the grace of Millard Filmore and head of the church of Mormon by direct succession. Difficulties after a while arose, however, between the Mormons and the Gentiles, the country was too narrow for both to live in. Mutual bigotry, begot mutual hatred; and the State of Deseret threatened to set up an independent Sovereignty in the Utah country. Popular clamor in the States demanded that this presumption should be punished and curbed, and the U. S. Government dispatched a formidable force under Gen. Harney to chastise them if need be, into subjection—Persifor F. Smith, commanding the Western department. In the summer of 1857, the army took up its line of march; but as they approached the confines of Mormondom, they were met with the white flag of peace, and though the difficulty is not yet arranged, it is not probable that any serious consequences will result from the Mormon war. The settlement at Salt Lake, even in its infancy, was regarded as a neucleus or rather as a point of departure for those interested in the Railroad enterprise, and was hailed as a fortunate event, being about equi-distant from the two extremities of the road, and near what was supposed to be the most eligible line. The question of a route, however, was yet in the dark, and promised to be the rock upon which the entire enterprise would split, unless managed with great prudence and circumspection. It was determined to have all such questions definitely settled by authority. Accordingly on the 3rd. March, 1853, Congress ordered to be made a series of explorations for the purpose of ascertaining the best and most economical route for such a railroad as was contemplated. The U. S. Topographical corps was called upon, and different surveying companies organized under command of Captain Pope, Captain Gunnison, Lieut.

Whipple, Lieut. Landor, and others to the number of some half dozen, and put upon the duty of a thorough and complete exploration. Belts of country, 200 miles in width, extending across the continent were assigned to each party, and all entered upon duty nearly at the same time. The result of their labors was a most complete and thorough report not only as to the topography; but the geology and botany of the country, together with minute descriptions of the animals and insects; and a complete classification as far as practicable, of the Indian tribes. Their report was published by authority of Congress and is a valuable addition to the literature of the country. Their explorations demonstrated the fact, that by more than one route it was practicable to construct railroads between the bounds designated in their instructions.—Gunnison's expedition which appears to have been successful in discovering the most eligible route, started from Fort Leavenworth in Kansas territory, in May 1853, followed the Missouri to the mouth of the Kansas and ascended it for a considerable distance along the usually travelled route of the Santa Fee traders, when it struck off in a South-west direction, their destination being the Huerfano river, in latitude about 38 deg. They passed through a country, hitherto almost unknown to the whites; inhabited by numerous Indians, prominent among whom were the Pah Utahs. On the Sevier river, about 150 miles from Salt Lake city, their camp was, on the 25th. October, 1853, surprised about daybreak by a band of these Indians, and Captain Gunnison and nearly his whole party massacred, before they could make resistance. The Mormons, were charged; but says the record of the expedition, unjustly, with inciting the massacre and through the exertions of Governor

Brigham Young, the papers, instruments and some of the horses were recovered from the Indians, and a head chief of the Pah Utahs explained by saying the murder was committed by some of the boys of the tribe in revenge for some of their fiends, whom they supposed had been killed by this party. He also, deprecated the vengeance of the whites, and promised to deliver up the murderers. The route by the Huerfano, had at this time been pronounced impracticable, being at the Pass of St. Luis, the dividing ridge between the Huerfano and the Rio Grande, 9,772 feet above the level of the sea, and the ascent being 1,118 feet in two and three quarter miles; but Captain E. G. Beckwith, having taken command of the expedition and reinforced the same, in a short time afterwards, near the Sierra San Juan they discovered a pass some 2000 feet lower, which was pronounced easily practicable. The waters of the Rio Grande del Norte, on the east, and those of the Rio Grande of the west, a branch of the western Colorado here interlock, not very far from Pike's Peak; the latter flowing into the Gulf of California, the former into the Gulf of Mexico.

Routes examined by other Engineers were pronounced more or less feasible; but this seems to have been the most practicable. The nearest approach to the old route of Lewis and Clarke, was one made by Mr. Landor which follows the Missouri nearly to its northern lend, crosses the mountain at Bridger's Pass, and then branches; one down the Sacramento to San Francisco, the other toward Puget's Sound by way of Lewis river, &c. This, is considered one of the best routes discovered, except the common objection of extreme cold. It is claimed however, that owing to peculiar circumstances, the passage of the mountains can be effected with less exposure to extreme cold, than by the more Southern routes. The expedition of Lewis and Clarke, did not experience any very extremely cold weather in this portion of their wanderings; though they experienced much inconvenience from the snow, and from utter ignorance of the country, Lewis and Clarke were forced to depend upon their own sagacity and to find their way almost unassisted through the trackless wilderness. As a matter of curiosity and reference we here insert from the Journal of Mr. Gass, "A memorandum of the computed distance in miles to the furthest point of discovery on the Pacific ocean, from the place where the canoes were deposited near the head of the Missouri, which from its mouth is,

IN MILES:	3096
From place of deposit to head spring,	24
To first fork of the Sho-sho-ne river,	14
To first large fork down the river,	18
To forks of the road at the mouth of Tour creek,	14
To fishing creek, after leaving the river,	23
To Flathead, or Clarke's river at Fish camp,	41
To the mouth of Travellers-rest creek,	76
To the foot of the great range of Mountains east side,	12
To the foot of the great range of Mountains west side,	130
To the Flathead village in a plain,	3
To the Koos-koos-ke river,	18
To the Canoe camp, at the forks,	6
To the Ki-moo-ee-nem,	60
To the Great Columbia, by Lewis' river,	140
To the mouth of the Sho-sho-ne, or Snake river,	162
To the Great Falls of Columbia,	6

To the Short Narrows,	3
To the Long Narrows	3
To the mouth of Catarack river, north side,	23
To the Grand Shoot, or Rapids,	42
To the Last Rapids, or Strawberry Island,	6
To the mouth of Quicksand river, south side,	26
To Shallow Bay at salt water,	136
To Blustry Point on North side,	13
To Point Open-slope, below encampment,	3
To Chin-Ooh river at bottom of Haley's Bay,	12
To Cape Disappointment on Western ocean,	13
To Capt. Clarke's tour N. W. along coast,	10
Total number of miles,	4133

These distances are of course only approximate, and not many of the names can be found on modern maps; yet they give an idea of the route traversed, that may be useful to understand properly the difficulties encountered. It is only marvellous that they made their escape, at all from the labyrinth of mountains and rivers in which they found themselves. The journals of late explorers, do not vary materially in the main features of their descriptions, from their accounts. It is apparent from the comparison, that forty years experience has not improved either the manners or the morals of the natives. On the contrary, they have not only become more immoral among themselves, but more disposed to be hostile toward the whites. Unprincipled white men have corrupted their morals, furnished them with whiskey, and rendered nugatory the well-meant endeavors of the U. S. Government, to ameliorate their condition. Of late years, the government has engaged zealously in the task of elevating them in the scale of civilization, and from the published reports of its agents, the effort has been attended with some success. Lieut. Whipple divides the Western Indians into three classes:--the semi civilized, the rude, and the barbarous. The first, comprise those who have been removed from the cast to the Mississippi, such as the

Choctaws	15,000
Chicasaws,	4,000
Cherokees,	17,000
Creeks and Seminoles,	24,000
Quapaws,	200
Shawnees,	300
Delawares,	250

making an aggregate of 62,000 persons, peaceful in their disposition and depending upon agriculture alone. They are characterized by docility and have a desire to learn and practice the manners, language and customs of the whites. The labors of missionaries among them, have been crowned with success and there appears to be no obstacle in the way, to prevent their complete civilization. The Shawnees and Delawares of this region do not participate in the favors bestowed upon the more northerly, bands of these tribes; and therefore complain that the Government overlooks their interests; as it bestows upon them neither annuities as to Choctaws, nor presents, such as are distributed among the wild tribes of the prairies. They evidently have an idea that the latter are given to the wild Indians as a kind of tribute, for fear of their depredations, and naturally murmur that they, who have always been friendly to the whites, should receive no assistance from them.

"Among those characterized as rude, may be enumerated the following, living in the Creek and Choctaw territories:

Toprofkies,	200
Kichais,	500
Kickapoos,	400
Caddoes,	100
Huecos,	400
Witchitas,	500

These remnants of tribes have much intercourse with, and are supposed to be considerably influenced by the semi-civilized class above alluded to. They cultivate the soil to some extent, but still retain many of their old habits, are fond of a roving life, and commit occasional depredations upon their neighbors.

The third class, denominated barbarous, are the Arabs of the plains, and the scourge of emigrants. According to the best information, their names and numbers are as follows:

Camanches,	20,000
Kaiowas,	3,500
Lipans,	6,500

amounting to about 30,000 persons, one fifth of whom are supposed to be warriors. They are perfect types of the American Savage and fully as barbarous as when first known to the Spaniards, centuries ago. They appear to be utterly irreclaimable either by kindness or force. From the earliest discovery of these tribes in the sixteenth century, they have preserved the same general character, that of an unconquerable indisposition to affiliate with the whites or in any manner to adopt their manners, customs or languages. A spirit of wild independence seems to possess them. They delight in rapine and make frequent incursions into the settlements of New Mexico, and are regarded by the more timid half breeds and Mexican Indians with the greatest fear. The appearance of a small band of Camanches, is sufficient to depopulate a whole village of these latter, and though they are somewhat wary in their collisions with the more energetic and warlike Texans, they not unfrequently make a foray upon the villages of that state and are off to their mountain fastnesses before pursuit can be hardly commenced. They have a wholesome respect however, for the *Americanos* of the North, which keeps them in some restraint.

The Kaiowas are kindred to the Camanches; and both are said to be branches of the Snake tribe, as is judged from their language and customs. The Lipans belong to the same general family, and are very numerous. Hunting and war are the favorite pursuits of these people. Agriculture is esteemed a degradation, from which their proud nature revolts, their dependence being upon game and depredations upon frontier settlements. So haughty is their spirit and so great their contempt for white men, that it is doubtful whether they will ever be induced to accept civilization and a local habitation; instead of the unrestrained freedom of their wild and savage life.

South and west of the Camanches, we come into the country of the Apaches, a people represented as more untamable even than the Camanches, to whom they also appear to be related. They cover a wide territory, and embrace some ten tribes, each of which governs itself independently; but recognizes a general bond of union. All these tribes acknowledge some sort of authority in the Spanish governors of New Mexico. The Navajos, the most northern of the Apache tribes, are, more given to settled habits than any of their congeners, and possess considerable flocks and herds. They are said to number about 8000 souls.

Both the Camanches and the Apaches are terrors to the more timid Indians and half breeds of New Mexico, and relying upon the terror with which they know themselves to be invested, they levy regular contributions upon their more indolent neighbors of the village and haciendas. Swooping down from their hills they spread terror and destruction in their paths. They are all expert horsemen, and though cruel, unscrupulous and bloodthirsty, are yet, not remarkably courageous notwithstanding their vain-glory and terrible reputation. They can be controlled by appealing to their fears and obtaining their respect by the certain conviction that depredations can and will be avenged. Since the establishment of American military posts in New Mexico, they have become much more tractable. In battle they are no match for the Texas rangers; a squad of whom, will put ten times their number of such Indians to flight. They generally, in such cases, depend much more upon stratagem than on valor. They are, moreover, considered faithless to their treaty obligations, when compelled to treat; and on the whole, are troublesome and very disagreeable neighbors.

Besides these more prominent Indian tribes, inhabiting the vallies of the Rio Grande, the Colorado and the Gila; there are numerous other minor tribes, with the same general characteristics, and of the same derivation, whose manners, customs, language and general characters have been ascertained and described, but for which we have no space.

One thing is apparent. Either the ancient Spanish travellers—Fathers Marco and Ruyz, Captains Alancon and Colonado, and others,—who wrote about the country of the Rio Grande as early as 1540, were very great romancers; or else, there have been exceedingly great changes wrought in the aspect of the country, and the character of its population, since their day. These writers all agree in their descriptions of an advanced state of civilization existing throughout this region; and in the country of Sevola or Cibola, they speak of having seen lofty houses built of stone, the people wearing dresses made of cotton, and living under good laws and regulations, that were as well observed as in civilized countries; and as being very numerous—in one province, alone, the population being 40,000 souls. There are many indications existing, of a state of civilization much more advanced than the present, having once prevailed throughout the region in question; but none to warrant any such representations as are made by these travellers. The country has evidently been once much more thickly settled, as appears from the numerous ruins, from these accounts, and from the traditions of the Indians themselves. It presents the rather singular appearance of a people in a state of active deterioration, from causes inherent among themselves; and at the same rate of diminution as has apparently prevailed among them since their first discovery, the present tribes will wear themselves out in a very few generations. The city of Zuni, is a type of these ancient cities of Cibola, several of which still actually exist, in ruins.

The Zuni district is situated between 32 and 35 deg. of north latitude and of longitude 108 and 113; and the city of the same name, is built up with long ranges of stone walls with an occasional opening near the top, for look out purposes. Entrance and egress is by means of ladders. It is still sparsely inhabited.

West from the Navajos, and in a fork between the little and the big Colorado, lies the country of the Moquinas, a people famous in Spanish history, as well for their devotion to liberty and successful valor in resisting foreign aggressions, as for their hospitality, integrity of character, and attention to agriculture. In many respects they assimilate to the people of Zuni, with whom they ever maintain friendly relations. The total population of the Moquinas is given at about 7000 and the tribe is spoken of as exceeding most of their neighbors in good qualities and energy of character.

To the north of the country inhabited by these tribes, is located the country of the Utahs, which is also a generic name, including several minor branches or tribes, acknowledging a common authority. The Pai-utes, or Pah-Utahs, of the vicinity of Great Salt Lake, are the most prominent among these tribes; and may be considered types for all of them. This tribe, however, it is said, does not number over 300 individuals, extremely vicious and very much disposed to be troublesome.

The total number of Indians living south of the Salt Lake route to California, and north of the present Mexican line, is estimated by Lieut. Whipple at 144,000; other authorities make it more or less, but this may be considered as approximating to the truth. Efforts have been made to systematize the languages of these Indians, and to trace some connection between the different families and tribes that are scattered over the vast area: but all such attempts are rather fanciful than valuable, and the surmises made, are much more curious than reliable. The Indians on the Colorado, are generally pronounced superior in all manly qualities, to those of any others in this section of the continent, and the Moquinas and Mojaves are especially complimented for their bravery, generosity, and kindly dispositions.

All these Indians have religious traditions and customs, more or less distinctly defined. The wilder the Indian--the less he has seen of white men--the more implicit, it is said, is his trust in the invisible Deity. From their unity of faith and similarity of modes of worship, Chisholm, an intelligent trader who resided many years among them, infers that the different tribes have all the same origin. The grand tenets of their belief are few, and very simple. They are: First--The existence of one Great Spirit. Second--A belief in future rewards, but not in future punishments. They have no idea of a hell, except what they have derived from the whites, believing that the wicked receive their deserts in this world, in sickness, poverty, war and death. Their modes of manifesting their belief are various, although there is much similarity among them, even in this respect. The Creeks worshipped fire, as the representative of purity and Deity, the Cherokees, and many other tribes had similar notions in regard to this element. The priestly office was widely recognized among all the tribes in the conjurations of the medicine man, and in some tribes, particular families were set apart and consecrated to the priesthood. They practice baptism and offer burnt sacrifices by way of thanksgiving or invocation. The number seven has a peculiar significance among many of the tribes; and indeed the points of contact in their beliefs and superstitions are so many and so decided, that the reader is irresistably forced, not only to the conviction that they are of a common stock, but that their beliefs have some connection with Mosaic revelation.

    The Pueblos Indians, say there is but one God and that *Montezuma,* a name of great repute among them, is his equal. Inferior to both is the sun, to whom they pray, because he looks upon them, knows their wants, and answers their prayers. The moon is younger sister to the sun and the stars are their children. Besides these, there is the Great Snake, to whom, by order of Montezuma, they are to look for life. These Indians, although nominally, professing Catholics, have in reality, little regard for the Catholic religion. In secret they glory in loyalty to Montezuma. They endeavor to keep their Spanish neighbors ignorant of their ceremonies; but they say, that Americans are brothers of the children of Montezuma, and their friends; therefore, they hide nothing from them. "Beneath," says our author, "the multiplicity of Gods, these Indians have a firm faith in the Deity, the unseen Spirit of Good. His name is above all things sacred, and, like Jehovah of the Jews, too holy to be spoken." The Apaches from superstitions reasons, will not kill or eat bears, and they have been known to refuse pork, even when suffering from hunger and when any less questionable food, however revolting in other respects, would have been eagerly eaten.

    As a general rule it may be stated that the farther

north, after leaving 30 degrees north latitude, we travel, the more the Indian character deteriorates, until it dwindles into the Esquimaux of Greenland and the Polar regions. The Indians of the Upper Missouri and the Columbia, encountered by Lewis and Clarke's expedition were generally inferior in body and mind to those farther to the South, some of whose characteristics we have been giving. In the main features, it is true, there is a decided similarity, sufficient to indicate a common origin; but there is lacking in the more northern Indians, the spirit of enterprise, of energy and springhtliness of intellect, that pertains to their more southern neighbors; and which in old times culminated in the semi-civilized communities of ancient Mexico. Whether, as they journeyed South into the more generous climate and soil of Mexico, the character of the aborigines was improved by natural causes, until they became builders of cities, instead of wanderers on the plains; or whether in their southern march the fierce savages of the north, met at the Colorado and the Rio Grande, the more gentle Aztecs, and blending with them formed a less polished, but a more vigorous race, is ground for a theory. Either, at least looks plausible. The grand question however, is not whence came the Indians; but whither go they? Since the time of our expedition, whiskey, the small pox and the cholera, have ravaged the numerous tribes passed by them on their route, until some of the most numerous bands have become almost extinct. This is the history of their friends the Mandans, who are now reported as numbering only 250 souls; as it is, more or less, of every tribe with whom the white man comes in contact.

The Rickarees, Snakes, Ponchas, Grosventres

and other tribes that were represented in 1805 as tolerably numerous and powerful, have dwindled until their numbers have become actually insignificant; while even the large and powerful family of the Sioux, at that day the most formidable Indians almost, known upon the plains, have fallen off in numbers until now, they scarcely boast a shadow of their ancient renown. It may be gratifying to know that these early acquaintances of our travellers, have since shown a commendable disposition to embrace civilization and agriculture and forsake their precarious and roving life of hunting and depredating upon their neighbors. The reports of the agents and missionaries show that they more willingly receive instruction than almost any other tribes of Indians whom it has been attempted to civilize. Schools are established among them and numerously attended by the youths of the tribes; while the elders in many cases have gone contentedly to work in splitting rails, ploughing their lands, and preparing themselves to become citizens of the United States.—There has been of late years an increase of attention bestowed upon all these tribes. During President Pierce's administration, alone, there were fifty-two separate Indian treaties made, and the Indian title to over 174,000,000 acres of land, peaceably relinquished into the hands of the government at a cost of about a quarter of a million of dollars. The Indian appropriations per year, in annuities, presents, salaries of Agents, &c., amount now to about one million dollars per year. The total number of Indians living within the limits of the United States and territories is given at about 350,000; of whom, about 150,000 inhabit New Mexico and the territory bordering thereon; some 60,000 the Missouri and branches; and the remainder are distributed over the Pacific slope of the continent from Puget's Sound to the southern extreme of California.

They are being gradually hemmed in on both sides, and the waves of white population will in a few years more meet in the midst of the plains; and the hunting grounds of the Indians will be known as separate possessions, no more, forever.

The following well written extract, we take from the report of Thomas S. Twiss, Indian agent on the upper Platte to the Commissioner of Indian affairs, Sept. 15th. 1856. The entire report is creditable alike to the heart and head of the agent, and if equally humane considerations as he evinces actuated more of our public men in regard to the Indians, there would be fewer difficulties.

"The wild Indian of the prairies is not very different from the wild Indian as described by the early colonists of the Atlantic States. The men are proud, haughty, independent, dignified in their bearing, observers of ceremony in their intercourse with the whites and with each other. They are taught to look upon manual labor as degrading and beneath the rank of the red man, whether he be chief, warrior, or brave. All menial services and labor are performed by the women, who are real slaves to the men. The only education of the latter is on the war path, and the only labor the pursuit of game. Beyond these he has no subjects of thought, or exercise for his mental faculties, and as a natural consequence, he is listless and idle during the greater part of his time.

"On the war path or in the chase he becomes intensely excited, and undergoes fatigue, and suffers for want of food, from cold and thirst, watches his enemy or his game, until he is certain of striking with deadly effect.

Then, when he returns to his lodge, he joins in the war dance, or in the feasts, and afterwards sinks into that apathy and indifference to all surrounding objects, which has so often been observed and commented upon by the whites, and which to them appears so strange and singular, that they judge, though erroneously, that the Indian is destitute of sensibility, feeling, or emotions.—Yet the reverse of this is the truth. There is not to be found among any people a more cheerful, contented and kindly disposed being than the Indian, when he is treated with kindness and humanity. His friendships are strong and lasting, and his love for and attachment to his children, kindred and tribe, have a depth and intensity which place him on an equality with the civilized race. His love and veneration for the whites amount to adoration, which is only changed to hatred and revenge by oppression, cruelties and deep wrongs and injuries inflicted upon the poor Indian, by the white man, without cause or reason. By his education on the war path, which leads to honor, fame and distinction, the Indian is a relentless, terrible enemy; he spares neither age nor sex, nor condition, but slaughters every one that comes in his path indiscriminately. He neither knows nor heeds the laws of modern warfare, as practiced and observed by an enlightened civilization. As a consequence, the first yell of the war whoop has scarcely died away in its distant echoes before a war of extermination is begun and waged against the poor Indian, and the innocent and the guilty alike perish, and their bones are left to bleach on their own happy hunting grounds. This is but a faint picture of Indian wars that have waged for short periods in every State and Territory in the Union, and which will burst forth constantly, until the power of the government is exerted to remove lawless and desperate whites from the Indian country, and change the habits of the Indian from a roving and hunter life to one of agriculture and fixed habitations.

"It may not be considered out of place, I trust, if I should state my opinions, formed from a careful observation and some experience as to the possibility of a combination or union of the wild tribes of the prairies, to wage war against the United States, which would necessarily be protracted and expensive. It would require a mighty genius to combine all the prairie tribes in hostility to the government. Such a genius must possess powers of oratory and persuasion, and far-seeing policy, and a popularity greater than that of a King Phillip, a Pontiac or a Tecumseh. If such a chief were to appear on the prairie now, he would find it a task of Herculean labor to form a party, the professed object of which should be hostility to the government. It would be an utter impossibility to harmonize discordant elements, smooth over difficulties, to heal old wounds and differences existing among the different tribes, or between bands of the same tribe. Besides, the chiefs are truly democratic, and are extremely jealous of each other, and it is not uncommon to hear that a particular chief has been deposed or passed over, because of his too great popularity, effected by a combination of petty chiefs, each of whom aspired to the office of head or principal chief. Other causes would render it a matter of great difficulty to unite different tribes, one of which is their own constant wars and feuds, which are unceasing; between whom there is never a peace nor even a truce.

"It would require the genius and military talents,

the powers of calculation and combination of a Napoleon, to form and maintain a union of these tribes for any length of time.

"If the reasons above stated are not sufficiently strong to prove that a combination of the wild tribes to wage a war against the authority of the government is utterly and absolutely impossible, for want of a master spirit, to unite, guide, and control them, and the chances of such a leader appearing upon the prairies being small, and even if he should make his advent, adverse circumstances are so many, and apparently insurmountable, that even momentary success could not be calculated upon, another and still stronger reason may be advanced, which is sufficient of itself, without any other, to settle this question of combination at once, and put it to rest forever. It is this: The Indians entertain no hostile or unfriendly feelings toward the government. It has not oppressed nor wronged them. They do not seek for any redress of grievances, either real or imaginary, for there are none. The parental care of the government to watch over their interests, to ameliorate their condition, to provide for their wants and necessities, and to protect them in their rights, is so plain and obvious to the Indians, that they see and feel, and express themselves on all occasions, that this guardianship is for their good and welfare, and the protection of the United States is the only shield by which they can hope for safety on the prairies, surrounded as they are on all sides by enemies. They make no complaints against any injustice or tyranny exercised toward them by official agents. It is only against those unprincipled whites who reside in their midst, in violation of law, that they complain of being wronged, cheated, insulted and beaten. It is certain from the most abundant evidence that the tribes, separately and collectively, are not disaffected to the government. They are friendly and well disposed, and desire to maintain their peaceful and amicable relations with it. This feeling of affection and gratitude to their "Great Father" is so strong and deep rooted that, it is not in the power of man to break or change it, except momentarily. Hence, if the proper and fit leader, should arise, yet it would be a task not easily accomplished of combining the tribes for an offensive or defensive war, consequently all cause of danger on that question may be dismissed, and we need apprehend none but outbreaks in which but a very small number of Indians of any one band is engaged in hostility.

"The Indians generally, and more especially the old chiefs and principal men, are shrewd and acute reasoners, considering that they have no advantages of education, no books of philosophy and history to guide them by the lights of truth and precedent. Their only history is oral tradition, mixed with much fable, handed down from generation to generation. As to the intellect, they are not deficient, and cannot be placed in a scale much below the white race, certainly not in a rank of great inferiority. The mind of the Indian lies a barren waste, without education, or training in processes of reasoning or logical deductions, except by such modes or paths as each one may happen accidentally to strike out for himself. Their amusements are few and simple; their virtues many; and vices were unknown among them until contaminated, debased and degraded by the white man. The old chiefs in council have often called my attention to their condition, and desired that I would request their "Great Father" to send them a farmer to teach the old men and women how to cultivate the earth, and raise corn for food; that they might, also have a teacher for the young children, and a missionary of the Gospel to teach them the ways of the Great Spirit. If our Great Father will be pleased to do these things for us, we shall have subjects of thoughts and attention to these things, and shall not think of going out upon the war trail. We shall stay at home and be quiet. We wish to be like the white man; to learn his ways of living, and, like him, to have subjects of thought and occupation. We see you, father, for days sitting in your lodge, and reading in the great book. We know that you are conversing with the Great Spirit, or with friends that live near the rising sun. You cannot see them, yet you are able to talk with them. We also see yon engaged in writing for many hours and know that you are talking to our "Great Father," and asking him to take pity on his red children. When thus occupied, you do not think about going upon the war path against your enemies; you are quiet and happy at home. We wish to be the same. We desire to be occupied with those things which are useful and necessary for us.—Now we have but little to amuse or occupy the mind. We are anxious to do good, and please our Great Father, but we often fail for want of judgment and forethought, which would not be the case if were educated and trained like the white man."

Such is a brief sketch of what has been brought to my notice and observation in my intercourse with the wild tribes of the prairies. I trust that the department will take such steps and adopt such measures, as in its judgment and wisdom may seem best to ameliorate and improve the condition of these poor Indians; to consider the plan of colonization, if that should be deemed a proper course to change them from a hunting to an agricultural people, or to carry into effect any other method that may be devised, in order to save these Indians from those wars of extermination which are invariably marked in their progress, by an indiscriminate slaughter of the innocent alike with the guilty, and the merciless and relentless massacre of unoffending women and helpless children."

In 1805, the country on both sides of the Mississippi, and the Illinois, the Wabash, the Lakes and even on the Ohio, was very similar in many respects to that now on the Missouri, the Yellowstone, the Kansas and the Platte; and, as we have seen large and populous commonwealths start up on these former rivers within that time, so we may reasonably expect in half a century more, the same development to take place in the latter. Already settlements have been pushed far up the Missouri. The Kansas country after being the scene of turmoil, confusion, political chicanery, and of actual warfare for a short period, has taken the initiatory steps for admission into the Union; and flourishing cities have sprung up, as if by incantation, where but a few short years ago, were but Indian lodges. Fort Leavenworth, the frontier post of a few dozen public buildings, of five years ago, has developed itself into a handsome city of some 6000 population; and the rolling plains, which Mr. Gass denominates as exceedingly beautiful and fertile, have been, acre after acre, appropriated; and are being rapidly dotted with the cabins of industrious settlers. This was not effected without exertion or without danger. The Slavery excitement, which had been revived in regard to Kansas, after the repeal of the Missouri Compromise, in 1850, by the desire of the

Missonrians, to make it a Slave State, and the opposition of the Northern people to such designs, became warm and active, about the year 1854. The act to organize the Territory, passed May 30th, 1854. At that time, there were but very few white residents in the Territory, though many were waiting for the Indian reservations to come into market, with the intention then of becoming settlers, or at least speculators. The Slavery controversy waxed warmer and hotter in Congress, and in the States; threatening even to divide the Union. The Missourians crossed the line and interfered in the territorial elections, Emigrant Aid Societies were organized in both sections, and the era of Sharpe's rifle, guerilla warfare, border ruffianism, anti-Slavery fanaticism and mutual outrage, was definitely inaugurated. Matters progressed, until the Territory was declared in a state of insurrection—the forms of law being disregarded by all parties, and the wildest anarchy prevailing. Brevet Major Gen. Persifer F. Smith, who died in 1858, at Leavenworth, universally lamented, was at that time Military commander of the Department. Governor after Governor:--Reeder, Shannon, Geary and Walker, were successively elected and deposed, or voluntarily resigned, in the short space of two years, being unable to enforce obedience or even command respect. In February, 1856, President Pierce declared by proclamation, martial law in the Territory, and the Military were ordered to restore order and enforce the laws of the United States. Rival legislative bodies had been organized, and capitals established at Topeka and Lecompton, fraud and violence prevailed and desperate remedies became imperative. The Topeka Convention was pronounced unlawful in its organization and objects; and on the 4th of July, 1856, Col. E.

V. Sumner, at the head of a body of U. S. Troops, forcibly dispersed it, while in session. Secretary of War, Jefferson Davis, censured him for this employment of the troops as unnecessary. Previously, on the 28th of June, Col. Jim Lane, having raised money and 250 volunteers, on the Free State side, threatened to march with large reinforcements, from Chicago across Iowa, to Council Bluffs. The pro-Slavery men, under Titus, Buford, Stringfellow, Atchison & Co., were also organized in numbers. The appearance of the Federal troops made the beligerents more wary, but did not stop the disturbance. Gen. Smith, on the 26th of July, announced the Territory peaceable; but in one month afterwards, 1200 armed men were said to be assembled at Lawrence, threatening to march over to Lecompton, the capital of the Territory, and destroy it; and on the authority of Gov. Shannon, Lieut. Col. Johnson was ordered to proceed to Lecompton, with all the troops at his disposal, to protect the public property. At this time the house of Col. Titus, near Lecompton, was attacked, one man killed, and several taken prisoners, who were afterwards delivered up to the Federal Authorities. The skirmish of Hickory Point, took place September 6th, between the Free Soilers, under Harvey, and the U. S. Troops, in which the former were surprised and dispersed.

Gov. Geary came into office, Sept. 11th, 1856, and issuing a conciliatory address, the war began to subside. From that date the reign of violence sensibly decreased, though Kansas affairs continued the staple of political discussion for a long time after, and were not finally determined until the decisive vote of the people of Kansas, in favor of a Free State policy, in August, 1858.

At this time the question may be considered as definitely settled, the Territory of Kansas has nearly sufficient population to qualify it for admission as a State, and before many more months she will emerge a fullfledged member of the American Union.

The Kansas controversy may now be considered closed, and the era of peace and good-will again prevails throughout the recently distracted Territory. A decided majority of the population appear to be opposed to Slavery as an institution of the prospective State; and the fact has induced a large immigration. Never in the history of the country have mens' minds been inflamed to so dangerous a degree, and never were the evils of sectional animosity so fearfully apparent. Nothing but the conservative good sense of the American character and the inherent sense of right of the American people warded off the imminent danger of a civil war, with the untold calamities, that must have followed in its train.

We have now brought our hero through a most important portion of his career, and sketched lightly the history of the Territory, of whose developement he and his party were the harbingers. Their memory still lingers among the natives of the Missouri, and the commanding officers have left their names indelibly impressed, not only upon the history but upon the geography of the country. So long as Lewis' river flows toward the Pacific, or the name of Clarke remains impressed upon the stream that bears his name, posterity will have them in perpetual remembrance. Nor have the subalterns been forgotten. In many an Indian lodge in the distant West, the old chiefs still speak of the white men who came among them first, with presents and the right hand of fellowship; and around the campfires of the trappers, as the Legends of the Missouri are told, the name of Sergeant Gass, is yet connected with many a tale of daring adventure.

Mr. Gass had now returned, in 1806, to the home of his friends, and acting under their advice, he resolved to have published the Journal of his travels. He had kept notes, in accordance with directions, during the entire expedition, but they were not exactly in a shape proper for publication, and his limited education precluding the idea of arranging them for the press, he secured the services of an Irish schoolmaster, named McKeehan. Mr. McKeehan undertook the task, and the result was an octavo volume, of 262 pages, which at the time, met with a great demand, principally owing to interest that was then felt in the subject, as the book itself, as it appears now, is rather dry, meagre end uninteresting. McKeehan presented his materials in the raw state, almost, and undigested, just as they were noted down by the author,--very frankly stating in his preface, that "neither he or Mr. Gass had attempted to give adequate representations of the scenes portrayed." Mr. Gass received the copy-right of the work, and one hundred copies of the first edition, while McKeehan re-ceived as his compensation, the balance of the edition, which he disposed of, to some profit. Mr. Gass realized but very little of the proceeds of his work, which appeared in the spring of 1807, printed by Zadoc Kramer, Pittsburgh. It has been since re-printed, in violation of his copy-right, and had some sale, as a matter of curiosity; but at this time it is out of print, and very few copies are believed to be in existence. The work lays no claim to beauty of diction, or much arrangement, but is valuable as being a correct, unvarnished record of the incidents of travel through an unknown region; and as showing Indian character in its true light, before being modified by intercourse with white men, and the vices of civilization. What the writings of Mother, Charlevoix and Smith are to the 'Salvages' of New England, Canada and Virginia, the Journal of Gass will be to the future historian, as to the aborigines of the future States some day to be organized on the banks of the Missouri, the Kansas, the Platte and the Columbia; with the difference in his favor, that his is free from the tendency to the marvellous, that so much distinguishes those veracious chroniclers.

But the excitement of authorship was too tame for our adventurous hero, and accordingly, the same spring of 1807, we find him again shaping his course for the frontier, and for the next four years of his life, he was engaged in various duties about the then outpost of Kaskaskia. For some time here, he held the post of assistant commissary, and transacted the duties of the office with his accustomed fidelity and zeal.

CHAPTER III.

THE WAR OF 1812.

1812, the smouldering embers of the Revolution broke out anew in the second war of Independence.— The overbearing conduct of the British officials and the tyrannical assumptions of their Government, had incensed the hate of the people of the United States, to the point of violence, notwithstanding that the country had but barely recovered from the exhaustion consequent upon the revolution; and was but poorly prepared for a long and arduous conflict with so powerful an enemy. At the conclusion of the revolutionary war, the British Government had acceded to the treaty recognizing the independence of the colonies, with a very bad grace, and up to the very time of signing the definitive article, of peace, Gen. Washington had been suspicious of treachery, and warned Congress and his countrymen, to be on their guard against the faithlessness of the British. It was the severest blow her pride had ever sustained. Although compelled to a formal recognition of independence, it was only after an eight years' war, after she had exhausted all her means in the colonies, had tired the patience of her people at home, and after she had been menaced by European combinations into the ungrateful necessity. She never cordially recognised the new republic as a member in the family of nations; although constrained by considerations of policy from any open demonstration of hostility against the rising commonwealths of her own blood and kin. The unsettled state of affairs in Europe during the latter part of the last century, and the wars of Napoleon that immediately followed in the beginning of the present, required all her attention for her own protection; but toward the close of this era, when the power of the great Corsican had begun to wane and she had a short respite from the machinations of her continental enemies, she turned her attention beyond the waters. The States had excited her resentment by affording aid and comfort to the French. She had never forgiven them for inviting and receiving assistance from France, during the revolutionary struggle; and when, during the almost utter annihilation of commerce between the European powers, consequent upon the wars, the Americans taking advantage of their neutrality, became the common carriers almost of the world, and did not discriminate against her enemy, the French; but were rather disposed to show them favor, her resentment was greatly increased. Assuming the sovereignty of the seas, she established an espionage of commerce. She asserted and exercised the right of search on the high seas, and condemned, without scruple, as contraband of war, whatever her officers saw fit to so declare. The merchant service was harrassed by her exactions, and American officers insulted on their own vessels. She claimed and exorcised the privilege of testing the nationality of the sailors on board the vessels boarded by her officers, and on the suspicion that they were subjects of Great Britain, she forcibly impressed into her service, from the decks of American merchantmen, not only foreign, but American born sailors. She affected a surveillance of our coasts, and in effect blockaded with her men-of-war our ports, so as seriously to impede commerce. She passed orders in council and executed them in defiance of our laws and remonstrances; and was rapidly vindicating by her actions her claim to be considered the mistress of the seas.

The Americans bore her insolent pretensions in no very patient spirit, for the old leaven of the revolution eras still fresh in the masses of the people; and many of the veterans of that contest were yet alive and on the stage of action. Yet they were slow to act. The country was just recovered from the depletion of the revolution, the continental wars had thrown an immense trade into the hands of our merchants and ship-owners, manufactures had begun to flourish, and the country was just starting upon the high road to prosperity. The industrial interests of the country demanded peace. It was well understood that the first hostile gun would be the signal for the swarming navy of Great Britain to pounce upon our scattered merchantmen, and sweep our commerce from the face of the ocean. Self-interest plead strongly for peace, even to the endurance of injury and insult. The merchants of the seaboard and their representatives in Congress, counseled forbearance; and as their interests were the interests that were most directly affected by the alledged causes of war, their remonstrances had great effect towards deferring the declaration of war. On the other hand, the national spirit of the country felt itself insulted,--pride was outraged, and from the interior, and from the distant west, came up deep and ominous sounds of dissatisfaction.—The war-spirit of the people was becoming aroused and the first notes of defiance came from the banks of the Ohio, the Mississippi and the Great Lakes. The western people were anxious for the war. At length, June 18th, 1812, during the Administration of Mr. Madison, war was formally declared. It was commenced without any exactly defined cause, and fought and ended, without the express recognition, by either party, of any contested question or principle. It appears to have been necessary, more to settle decisively and forever, questions of feeling than of policy; though of these latter, there were many that required settlement. These remained undecided for nearly half a century afterwards, when in 1858, the British handsomely and unequivocally renounced the right of search and impressment, which they had refused, at the point of the bayonet, in 1815. Still, although the war was apparently fruitless of consequences, it was practically final as to the grievances of which the Americans complained. The British Government, while it avoided any acknowledgement of the American doctrines regarding impressment, and the right of search, nevertheless, of their own accord, carefully avoided any repetition of the offence itself, confident that if they repeated it, they would have the war to fight over again; and that the indulgence would not pay them for the trouble it would cost. On this principle, although the right was never in so many words relinquished, no effort at exercising it was ever made, until that of 1858, which resulted so happily. This attempt to revive it, it may be proper to say, was ordered through a misconstruction of the sentiments of the American Secretary of State, and executed in another misconstruction of orders, by the officers in command of the British squadron, effecting the end it accomplished, much more by good luck than by good management.

The zeal with which the declaration of war was hailed in the Southern and Western portions of the Union is historical; as is also the reluctance of the Eastern and New England States to engage in it, to the detriment of their trading interests, and the unprepared state of the country at large, to enter upon along and arduous war, with their powerful and implacable foe. Still the news spread over the country like wild-fire, and was, in the West, hailed with enthusiasm. Volunteers crowded round the standard of the stars and stripes. The excitement reached the frontier, and a call was published at Kaskaskia, appealing to the patriotism of the pioneers to organize in the common defence. A company of rangers was quickly rendezvoused at Herculaneum, in Missouri, in consequence of the call. Mr. Gass was prevented, accidently, from joining this party, the organization and objects of which, suited his disposition, but quickly found that he would be needed in another capacity. For a short time previous to the declaration of war, he had been engaged in the lead trade—then thriving in the vicinity in which he was located, and it so happened that at the time the rangers were being organized, he was absent at Nashville, Tenn., with a quantity of lead, for the purpose of trade. While here, a great excitement existed in reference to the depredations of the Southern Indians,—men were being raised, and preparations made for a campaign against them; and he was, much against his will, drafted into the regiment raised by

General Jackson, to fight against the Creeks. He had the option, however, of enlisting in the regular army for five years, with $100 cash in advance, and a promise of $24 extra, on the expiration of his time of service and, perhaps coming to the conclusion that the line of his usefulness lay in a military direction, he forthwith enlisted for the war, under Gen. Gaines, and was immediately marched to the North, leaving his lead speculation in the hands of his partner.

At this time he came into contact with many of the military characters, then obscure, but who since have distinguished themselves in the annals of the country as soldiers and statesmen. To recapitulate the stirring scenes of that era is not within the province of our sketch, though as Mr. Gass, served through the three several campaigns faithfully in the service of his country, a cursory review of the war of 1813, might seem necessary to a proper elucidation of his character. That, however is the duty of the general historian and to repeat it here, would transcend the limits of our plan. In 1813, Mr. Gass, was stationed at Fort Massac, in Illinois, where he remained for some months occupied diligently in frontier duties, conciliating as far as possible the good will of the Indians with whom the British were at the same time tampering; and whom they were endeavoring to induce to take up arms against the Americans. This, with other such important, though unobtrusive services occupied his time at this period. Removing thence, to Bellfontaine and afterwards accompanied by a detachment of forty men, some forty miles above the mouth of the Illinois, they constructed there a fort. At this place, Mr. Gass, was so unfortunate as to lose an eye by being struck with a splinter from a falling tree.—The surgery at hand was very indifferent; and his eye healed up with difficulty, disabling him from active service for several months. Nursing his hurt through the winter of 1813-4; in the Spring, orders came to prepare with all speed a fleet of boats on the banks of the Illinois, in which his corps should proceed with all their munitions, to Pittsburgh--there to join the Northern army, that was assembling for the defence of the Canada frontier. A few weeks, sufficed to enable them to supply themselves with floating crafts and leaving their encampment, they embarked on the Illinois, down which they floated, until they struck the Mississippi, and thence down to the mouth of the Ohio. At this time the rivers were swollen with the spring rains and the low lands at the mouth of the Ohio were covered with water.—The current was swift and they had no means of ascending except by dint of pushing; and pulling by the trees on the banks. However, they persevered, and after labor they reached about the 1st. of July, 1814, Pittsburgh, their place of destination—having traversed in this laborious style, the whole length of the Ohio, from its mouth to its very head.

Arriving at Pittsburg, the men were at once formed into four companies under the immediate command of Col. Nicholls, and attached to the northern army commanded by Gen. Brown.

The war by this time had progressed and many important events had taken place. Many gallant encounters had taken place on the seas, in which the American arms were often victorious, demonstrating their ability to cope successfully with England on her favorite element; merchantmen, had been converted into privateers, and carried havoc among the mercantile marine of the enemy; and in the marine department of the war, the American arms were triumphant. On land, they were not so fortunate. At the outset of the war, the regular army was next to nothing in numbers; and although the President was authorized to call out 100,000 militia, experience soon demonstrated, that however patriotic the militia, and however brave in defence of their firesides, they were very unreliable in an aggressive war as this was in some respects to be. The militia, could with difficulty, often, be induced to march beyond the borders of their own States; and absolutely refused to cross into Canada, when the reduction of Canada became an object with the Americans. It required time, to organise a regularly drilled army, and consequently, the first campaigns of the war were anything but flattering to the prowess of the American arms. Gen. Hull, a veteran officer of the revolution, had surrendered his army at Detroit, in August 1813; Gen. Winchester had been defeated on the frontier in January, 1814, and his whole force compelled to capitulate to the British under Col. Proctor. The militia were generally uncontrollable and defeat and disaster seemed to be the order of the day. The theatre of war had become located on the Canada frontier. The Americans had attempted an invasion under Gen. Van Rensaleer and were successful, until thwarted by the "constitutional scruples" of the militia, and compelled to retreat. The British and Indians were in force along the whole line under Proctor, Riall and others; and the inhabitants were greatly harrassed by their constant forays. The lakes themselves were also the scenes of several gallant encounters between the American and English vessels, which resulted gloriously to the former. This frontier was the scene of more hard fighting than any other portion of the country. In the Spring of 1814, it was determined to make a desperate effort not only to retrieve the honors of the American arms by a brilliant campaign but to make a decisive strike toward the invasion and capture of Canada. By this time the Americans had become more accustomed to the service, and a better spirit had begun to actuate the militia. The army was in better drill, better organized and more efficiently officered than it had previously been. Men were ordered from Kentucky, Ohio and other western States to rendezvous at different points on the frontier; and in pursuance of this order, the detachment to which our hero was attached was forthwith marched from Pittsburgh, up French Creek, to Presque Isle, now Erie, and crossed over into Canada. Here, after a series of marches and counter-marches, without any actual collision with the enemy, although often in their immediate proximity, the detachment spent a few days in Canada, then re-crossed the Niagara at Black Rock into the States and in a very few days after, crossed again into Canada at Chippewa Fort. The American army at this time, was in two divisions--one commanded by Gen. Brown, the other by Gen. Macomb, both co-operating together.

Nile's Register gives the following account of the battle at Chippewa, to participate in which, Mr. Gass arrived but a few hours too late.

"On the evening of the 2nd of July, general orders were issued for the embarkation of the troops by daylight next morning, when the army, consisting of two brigades, and a body of New York and Pennsylvania volunteers and Indians, under General P. B. Porter, were landed on the opposite shore, without opposition. The first brigade, under Gen. Scott, and the artillery corps, under Major Hindman, landed nearly a mile below Fort Erie, while General Ripley, with the second made the shore about the same distance above. The fort was soon completely invested, and a battery of long eighteens being planted in a position which commanded it, the garrison, consisting of 137 men, including officers, surrendered prisoners of war. Several pieces of ordnance were found in the fort, and some military stores.

Having placed a small garrison in Fort Erie to secure his rear, Brown moved forward the following day to Chippewa plains, where he encamped for the night, after some skirmishes with the enemy.

The American pickets were several times attaked on the morning of the 5th, by small parties of the British. About four in the afternoon, General Porter, with the volunteers and Indians, was ordered to advance from the rear of the American camp, and take a circuit through the woods to the left, in hopes of getting beyond the skirmishing parties of the enemy, and cutting off their retreat, and to favor this purpose the advance were ordered to fall back gradually under the enemy's fire. In about half an hour, however, Porter's advance met the light parties in the woods, and drove them until the whole column of the British was met in order of battle. From the clouds of dust and the heavy firing, General Brown concluded that the entire force of the British was in motion, and instantly gave orders for General Scott to advance with his brigade and Towson's artillery, and meet them on the plain in front of the American camp. In a few minutes Scott was in close action with a superior force of British regulars.

By this time, Porter's volunteers having given way and fled, the left flank of Scott's brigade became much exposed. General Ripley, was accordingly ordered to advance with a part of the reserve, and skirting the woods on the left, in order to keep out of view, endeavor to gain the rear of the enemy's right flank. The greatest exertions were made to gain it, but in vain.—Such was the gallantry and impetuosity of the brigade of General Scott, that its advance upon the enemy was not to be checked. Major Jessup, commanding the batallion on the left flank, finding himself pressed both in front and in flank, and his men falling around him, ordered his batallion to "support arms and advance."—Amidst the most destructive fire this order was promptly obeyed, and he soon gained a more secure position, and returned upon the enemy so galling a discharge, as caused them to retire.

The whole line of the British now fell back, and the American troops closely pressed upon them. As soon as the former gained the sloping ground descending towards Chippewa, they broke and ran to their works, distant about a quarter of a mile, and the batteries opening on the American line, considerably checked the pursuit. Brown now ordered the ordnance to be brought up, with the intention of forcing the works. But on their being examined, he was induced by the lateness of the hour, and the advice of his officer, to order the forces to retire to camp.

"The American official account states their loss at 60 killed, 248 wounded, and 19 missing. The British officially state theirs at 132 killed, 320 wounded, and 46 missing.

"Dispirited as was the public mind at this period, the intelligence of this brilliant and unexpected opening of the campaign on the Niagara could not fail of being most joyously received. The total overthrow of the French power had a few months before liberated the whole of the British forces in Europe. A considerable portion of Lord Wellington's army, flushed with their late success in Spain, had arrived in Canada, and were actually opposed to Brown at Chippewa, while all our maritime towns were threatened by Britain's victorious armies, whose arrival was momentarily expected on the coast. When the intelligence of the stupendous events in Europe was first received, many consoled themselves with the idea, that the magnanimity of Great Britain would freely grant in her prosperity, what they insisted we never could force from her in her adversity. Sincerely taking for realities the pretexts on which our neutral rights had been infringed, they thought the question of impressment, now the almost single subject of dispute, could easily be amicably arranged, when the affairs of the world were so altered as to render it nearly impossible that Great Britain could ever again be reduced to the necessity of "fighting foe her existence;" or, at all events, as the peace of Europe had effectually removed the cause, and as the American government declined insisting on a formal relinquishment of the practice, no difficulty would be thrown in the way of a general and complete pacification of the world.

"This illusion was soon dissipated. By the next advices from Europe it was learned, that the cry for vengeance upon the Americans was almost unanimous throughout the British empire. The president was threatened with the fate of Bonaparte, and it was said that the American peace ought to be dictated in Washington, as that of Europe had beers at Paris. Even in parliament the idea was held out that peace ought not to be thought of till America lead received a signal punishment, for having dared to declare war upon them while their forces were engaged in "delivering Europe" from its oppressor. The commencement of the negotiations for peace, which had been proposed by the British court, was suspended, and strenuous efforts were made to send to America as commanding a force as possible.

"Under these circumstances, a victory gained by the raw troops of America over the veterans of Wellington, superior in numbers to the victors, upon an open plain, and upon a spot chosen by the British general, had a most beneficial tendency, by dispelling the dread which the prowess of the British troops in Spain could not have failed to have produced in the minds of their opponents. This battle was to the army what the victory of Captain Hull had been to the navy; and the confidence which it inspired was surely most justly founded, for every man felt that the victory had been gained by superior skill and discipline: it was not the fruit of any accidental mistake or confusion in the army of the enemy, or of one of those movements of temporary panic on one side, or excitement on the other, which sometimes gives a victory to irregular courage over veteran and disciplined valour.

"After so signal a defeat, the British could not be induced to hazzard another engagement. They abandoned their works at Chippewa, and burning their barracks, retired to Fort Niagara and fort George, closely followed by Brown. Here he expected to receive some heavy guns; and reinforcements from Sacketts's Harbour; but on the 23rd, of July he received a letter by express from general Gaines, advising him that that port was blockaded by a superior British force, and that Commodore Chauncy was confined to his bed with a fever. Thus disappointed in his expectations of being enabled to reduce the forts at the mouth of the Niagara, Brown determined to disencumber the army of baggage, and march directly for Burlington Heights. To mask this intention, and to draw from Schlosser a small supply of provisions, he fell back upon Chippewa.

"About noon on the 25th, general Brown was advised by an express from Lewistown, that the British were following him, and were in considerable force in Queens-town and on its heights, that four of the enemies fleet had arrived with reinforcements at Niagara during the preceding night, and that a number of boats were in view, moving up the river. Shortly after, intelligence was brought that the enemy were landing at Lewistown, and that the baggage and stores at Schlosser, and on their way thither, were in danger of immediate capture. In order to recall the British from this object, Brown determined to put the army in motion towards Queenstown, and accordingly General Scott was directed to advance with the first brigade, Towson's artillery, and all the dragoons and mounted men, with orders to report if the enemy appeared, and if necessary to call for assistance. On his arrival near the Falls, Scott learned that the enemy was in force directly in his front, a narrow piece of woods alone intercepting his view of them. He immediately advanced upon them, after dispatching a messenger to General Brown with this intelligence.

The reports of the cannon reached General Brown before the messenger, and orders were instantly issued for General Ripley to march to the support of General

Scott, with the second brigade and all the artillery; and Brown himself, repaired with all speed to the scene of action, whence he sent orders for General Porter to advance with his volunteers. On reaching the field of battle, General Brown found that Scott had passed the wood, and engaged the enemy on the Queenstown road and on the ground to the left of it, with the 9th, 11th and 22nd regiments, and Towson's artillery, the 25th , having been thrown to the right to be governed by circumstances. The contest was close and desperate, and the American troops, far inferior in numbers, suffered severely.

"Meanwhile, Major Jessup, who commanded the 25th regiment, taking advantage of a fault committed by the British commander, by leaving a road unguarded on his left, threw himself promptly into the rear of the enemy, where he was enabled to operate with the happiest effect. The slaughter was dreadful; the enemy's line fled down the road at the third or fourth fire. The capture of Gen. Riall with a large escort of officers of rank, was part of the trophies of Jessup's intrepidity and skill; and, but for the impression of an unfounded report, under which he unfortunately remained for a few minutes; Lieutenant General Drummond, the commander of the British forces, would inevitably have fallen into his hands, an event which would, in all probability, have completed the disaster of the British army. Drummond was completely in Jessup's power; but being confidently informed that the first brigade was cut an pieces, and finding himself with less than 300 men, and without any prospect of support, in the midst of an overwhelming hostile force, he thought of nothing for the moment, but to make good his retreat, and save his command. Of this temporary suspense of the advance of the American column, General Drummond availed himself to make his escape. Among the officers captured, was one of General Drummond's aids-de-camp, who had been dispatched from the front line to order up the reserve, with a view to fall on Scott with the concentrated force of the whole army, and overwhelm him at a single effort. Nor would it have been possible to prevent this catastrophe, had the reserve arrived in time; the force with which General Scott would then have been obliged to contend being nearly quadruple that of his own. By the fortunate capture, however, of the British aid-de-camp, before the completion of the service on which he was ordered, the reserve was not brought into action until the arrival of General Ripley's brigade, which prevented the disaster which must otherwise have ensued.

"Though the second brigade pressed forward with the greatest ardour, the battle had raged for an hour before it could arrive on the field, by which time it was nearly dark. The enemy fell back on its approach. In order to disengage the exhausted troops of the first brigade, the fresh troops were ordered to pass Scott's line, and display in front, a movement which was immediately executed by Ripley. Meanwhile the enemy, being reconnoitered, was found to have taken a new position, and occupied a height with his artillery, supported by a line of infantry, which gave him great advantage, it being the key to the whole position. To secure the victory, it was necessary to carry his artillery and seize the height. For this purpose the second brigade advanced upon the Queenstown road, and the first regiment of infantry, which had arrived that day, and was attached to neither of the brigades, was formed in a line facing the enemy's on the height, with a view of drawing his fire and attracting his attention, as the second brigade advanced on his left flank to carry his artillery.

As soon as the first regiment approached its position, colonel Miller was ordered to advance with the 21st regiment, and carry the artillery on the height, with the bayonet. The first regiment gave way under the fire of the enemy; but Miller, undaunted by this occurrence, advanced steadily and gallantly to his object, and carried the heights and cannon in a masterly style. General Ripley followed on the right with the 23d regiment. It had some desperate fighting, which caused it to falter, but it was promptly rallied, and brought up.

"The enemy being now driven from their commanding ground, the whole brigade with the volunteers and artillery, and the first regiment, which had been rallied, were formed in line, with the captured cannon, nine pieces in the rear. Here they were soon joined by Maj. Jessup, with the 25th, the regiment that had acted with such effect in the rear of the enemy's left. In this situation the American troops withstood three distinct desperate attacks of the enemy, who had rallied his broken corps, and received reinforcements. In each of them he was repulsed with great slaughter, so near being his approach, that the buttons of the men were distinctly seen through the darkness by the flash of the muskets, and many prisoners were taken at the point of the bayonet, principally by Porter's volunteers. During the second attack General Scott was ordered up who had been held in reserve with three of his battalions, from the moment of Ripley's arrival on the field. During the third effort of the enemy, the direction of Scott's column would have enabled him in a few minutes, to have formed line in the rear of the enemy's right, and thus have brought him between two fires.--But a flank fire from a concealed party of the enemy falling upon the centre of Scott's command, completely frustrated this intention. His column was severed in two; one part passing to the rear, the other by the right flank of platoons towards Ripley's main line.

"This was the last effort of the British to regain their position and artillery, the American troops being left in quiet possession of the field. It was now nearly midnight, and Generals Brown and Scott being both severely wounded, and all the troops much exhausted, the command was given to General Ripley, and he was instructed to return to camp, bringing with him the wounded and the artillery.

"Gen. Ripley has been much blamed for the non-execution of this order, by which the captured cannon again fell into the hands of the British. Gen. Brown, in his official report says, 'To this order he (Ripley) made no objection, and I relied upon its execution.—It was not executed.' On the part of Gen. Ripley it is stated, that his orders were, in case an enemy appeared in force, 'to be governed entirely by circumstances.' His orders, therefore, were executed. At daybreak the army was arranged and the march commenced, when circumstances of the most positive nature were made apparent, such as must have been in view in the discretionary part of the order, and in the full effect of which General Ripley commenced and effected the retreat which afterwards led him to Fort Erie. The troops, reduced to less than 1600 men, were marched on the 26th by Gen. Ripley toward the field of battle.

Motion was commenced at day-break, but difficulties incidental to the late losses prevented the advance before some time had been spent in reorganization and arrangement. The line of march being assumed and the Chippewa crossed, Gen. Ripley sent forward Lieutenants Tappan of the 23d, and Riddle of the 15th, with their respective commands, to reconnoitre the enemy's position, strength and movements. On examination, he was found in advance of his former position, on an eminence, strongly reinforced, as had been asserted by prisoners taken the preceding evening; his flanks resting on a wood one side, and on the river on the other, defied being turned or driven in; his artillery was planted so as to sweep the road; besides these advantages, he extended a line nearly double in length to that which could be displayed by our troops. To attack with two-thirds the force of the preceding evening an enemy thus increased, was an act of madness that the first thought rejected. The army was kept in the field and in motion long enough to be assured of the strength and position of the enemy; that information being confirmed, there remained but one course to prevent that enemy from impeding a retreat, which, had he been vigilant, he would previously have prevented. The army, therefore, immediately retrogaded, and the retreat received the sanction of General Brown, previous to his crossing the Niagara.

The American official account states their loss in this battle at 171 killed, 572 wounded, and 117 missing; the return of British prisoners presents an aggregate of 169, including Major General Riall, and a number of officers. The British state their loss to be 84 killed, 559 wounded, 193 missing; their loss in prisoners they stated at only 41. Major-General Brown and Brigadier General Scott were among the wounded of the Americans, and Lieutenant General Drummond and Major General Riall among those of the British.

On the arrival of the British before Fort Erie, they perceived that the opportunity was lost of carrying the American works by a coup-de-main. Driving in the pickets, therefore, they made a regular investment of the place. The following day, General Gaines arrived from Sackett's Harbor, and being senior in rank assumed the command. On the 6th., the rifle corps was sent to endeavor to draw out the enemy, in order to try his strength. Their orders were, to pass through the intervening woods, to amuse the British light troops until their strong columns should get in motion, when they should retire slowly to the plain, where a strong line was posted in readiness to receive the enemy. The riflemen accordingly met and drove the light troops into their lines, but although they kept the wood nearly two hours, they were unable to draw any part of the enemy's force after them. The British left eleven killed and three prisoners in the hands of the riflemen; but their loss was supposed to be much more considerable. The loss of the riflemen was five killed and three or four wounded.

The main camp of the British was planted about two miles distant. In front of it, they threw up a partial circumvallation, extending around the American fortifications. This consisted of two lines of entrenchment, supported by block-houses; in front of these, at favorable points, batteries were erected, one of which enfiladed the American works.

"The American position was on the margin of

lake Erie, at the entrance of the Niagara river, on nearly horizontal plain, twelve or fifteen feet above the surface of the water, possessing few natural advantages. It had been strengthened in front by temporary parapet breast-works, entrenchments, and abbatis, with two batteries and six field-pieces. The small unfinished fort, Erie, with a 24, 18, and 12 pounder, formed the northeast, and the Douglas battery, with an 18 and 6 pounder near the edge of the lake, the south-east angle on the right. The left was defended by a redoubt battery with six field-pieces, just thrown up on a small ridge. The rear was left open to the lake, bordered by a rocky shore of easy ascent. The battery on the left was defended by Captain Towson; fort Erie by Captain Williams, with Major Trimble's command of the 19th infantry; the batteries on the front by Captains Biddle and Fanning; the whole of the artillery commanded by Major Hindman. Parts of the 9th, 11th, and 22d infantry, were posted on the right under the command of Lieutenant-colonel Aspinwall. General Ripley's brigade, consisting of the 21st and 23d, defended the left. General Porter's brigade of New York and Pennsylvania volunteers, with the riflemen, occupied the centre.

"During the 13th and 14th, the enemy kept up a brisk cannonade, which was sharply returned from the American batteries, without any considerable loss. One of their shells lodged in a small magazine, in fort Erie, which was almost empty. It blew up with an explosion more awful in appearance than injurious in its effects, as it did not disable a man or derange a gun. A momentary cessation of the thunders of the artillery took place on both sides. This was followed by a loud and joyous shout by the British army, which was instantly returned on the part of the Americans, who, amidst the smoke of the explosion, renewed the contest by an animated roar of the heavy cannon.

"From the supposed loss of ammunition, and the consequent depression such an event was likely to produce, General Gaines felt persuaded that this explosion would lead the enemy to assault, and made his arrangements accordingly. These suspicions were fully verified, by an attack that was made in the night between the 14th and 15th of August.

"The night was dark, and the early part of it raining, but nevertheless one third of the troops were kept at their posts. At half past two o'clock, the right column of the enemy approached, and though enveloped in darkness, was distinctly heard on the American left, and promptly marked by the musketry under majors Wood and captain Towson. Being mounted at the moment, Gaines repaired to the point of attack, where the sheet of fire rolling from Towson's battery, and the musketry of the left wing, enabled him to see the enemy's column of about 1500 men approaching on that point; his advance was not checked until it had approached within ten feet of the infantry. A line of loose brush, representing an abattis, only intervened; a column of the enemy attempted to pass round the abattis, through the water, where it was nearly breast-deep.—Apprehending that this point would be carried, Gaines ordered a detachment of riflemen and infantry to its support, but at this moment the enemy were repulsed. They instantly renewed the charge, and were again driven back.

"On the right, the fire of cannon and musketry announced the approach of the centre and left columns

of the enemy, under Colonels Drummond and Scott.—
The latter was received and repulsed by the 9th, under
the command of Captain Foster, and Captains Boughton
and Harding's companies of New York and Pennsylvania
volunteers, aided by a six pounder, judiciously posted by
Major M'Kee, chief engineer.

But the centre, led by Colonel Drummond, was
not long kept in check; it approached at once every assailable point of the fort, and with scaling ladders ascended
the parapet, where, however, it was repulsed with dreadful carnage. The assault was twice repeated and as often
checked; but the enemy having moved around in the ditch,
covered by darkness, increased by the heavy cloud of
smoke which had rolled from the cannon and musketry,
repeated the charge, re-ascended the ladders, and with
their pikes, bayonets and spears fell upon the American
artillerists, and succeeded in capturing the bastion. Lieutenant M'Donough, being severely wounded, demanded
quarter. It was refused by Colonel Drummond. The Lieutenant then seized a handspike, and nobly defended himself until he was shot down with a pistol by the monster
who had refused him quarter, who often reiterated the
order—"give the damned Yankees no quarter." This officer, whose bravery, if it had been seasoned with virtue,
would have entitled him to the admiration of every soldier—this hardened murderer soon met his fate. He was
shot through the breast, while repeating the order "to give
no quarter."

Several gallant attempts were made to recover
the right bastion, but all proved unsuccessful. At this
moment every operation was arrested by the explosion of
some cartridges deposited in the end of the stone building

adjoining the contested bastion. The explosion was tremendous and decisive; the bastion was restored by the flight of the British. At this moment Captain Biddle was ordered to cause a field piece to be posted so as to enfilade the exterior plain and salient glacis. Though not recovered from a severe contusion in the shoulder, received from one of the enemy's shells, Biddle promptly took his position, and served his fieldpiece with vivacity and effect. Captain Fannings battery likewise played upon them at this time with great effect. The enemy were in a few momonts entirely defeated, taken or put to flight, leaving on the field 221 killed, 174 wounded, and 186 prisoner, including 14 officers killed and 7 wounded and prisoners. A large portion were severely wounded; the slightly wounded, it is presumed were carried off.

The loss of the Americans during the assault was seventeen killed, fifty-six wounded, and 11 missing. The British accounts acknowledge only 57 killed, 309 wounded, and 539 missing. During the preceding bombardment, the loss of the Americans was 7 killed, 19 severely and 17 slightly wounded. The loss of the British is not mentioned in their official account. This bombardment commenced at sun-rise on the morning of the 13th, and continued without intermission till 8 o'clock P. M.; recommenced on the 14th, at day light, with increased warmth; and did not end until all hour before commencement of the assault on the morning of the 15th.

A short time after the assault on Fort Erie, General Gaines received a serious wound from the bursting of a shell, by which means the command once more devolved on General Ripley, till the 2d of September, when the state of his health allowed Gen. Brown again to place himself at the head of his army.

The troops in Fort Erie began now to be generally considered as in a critical situation, and much solicitude to be expressed for the fate of the army that had thrown so much glory on the American name, menaced as it was in front by an enemy of superior force, whose numbers were constantly receiving additions and whose batteries every day becoming more formidable, while a river of difficult passage lay on their rear. Reinforcements were ordered on from Champlain, but they were yet far distant. But the genius of Brown was fully equal to the contingency, and the difficulties with which he was environed served only to add to the number of his laurels.

"Though frequent skirmishes occurred about this period, in which individual gallantry was amply displayed, yet no event of material consequence took place till the 17th of September, when having suffered much from the fire of the enemy's batteries, and aware that a new one was about to be opened, General Brown resolved on a sortie in order to effect their destruction.—The British Infantry at this time consisted of three brigades, of 12 or 1500 men each, one of which was stationed at the works in front of Fort Erie, the other two occupied their camp behind. Brown's intention therefore was, to storm the batteries, destroy the cannon, and roughly handle the brigade upon duty, before those in reserve could be brought into action.

"On the morning of the 17th, the infantry and riflemen, regulars and militia, were ordered to be paraded and put in readiness to march precisely at 12 o'clock. General Porter with the volunteers, Colonel Gibson with the riflemen, and Major Brooks with the 23d and 1st infantry, and a few dragoons acting as infantry, were ordered to move from the extreme left upon the enemy's right, by a passage opened through the woods for the occasion. General Miller was directed to station his command in the ravine between Fort Erie and the enemy's batteries, by passing them by detachment through the skirts of the wood--and the 21st infantry under General Ripley was posted as a corps of reserve between the new bastions of Fort Erie--all under cover, and out of the view of the enemy.

"The left column, under the command of General Porter, which was destined to turn the enemy's right, having arrived near the British intrenchments, were ordered to advance and commence the action. Passing down the ravine, Brown, judged from the report of the musketry that the action had commenced. Hastening, therefore, to General Miller, he directed him to seize the moment and pierce the enemy's intrenchments between batteries No. 2 and 3. These orders were promptly and ably executed. Within 30 minutes after the first gun was fired, batteries No. 2 and 3, the enemy's line of entrenchments, and his two block-houses, were in possession of the Americans.

Soon after, battery No. 1 was abandoned by the British. The guns were then spiked or otherwise destroyed and the magazine of No. 3 was blown up.

A few minutes before the explosion, the reserve had been ordered up under General Ripley, and as soon as he arrived on the ground, he was ordered to strengthen the front line, which was then engaged with the enemy in order to protect the detachments employed in demolishing the captured works. While forming arrangements for acting on the enemy's camp during the moment of panic, Ripley received a severe wound. By this time, however, the object of the sortie being accomplished beyond the most sanguine expectations, General Miller had ordered the troops on the right to fall back, and observing this movement, Brown sent his staff along the line to call in the other corps. Within a few minutes they retired from the ravine, and thence to the camp.

"Thus, says General Brown, in his dispatch, 1,000 regulars and an equal portion of militia, in one hour of close action, blasted the hopes of the enemy, destroyed the fruits of fifty day's labor, and diminished his effective force 1000 men at least.

"In their official account of this sortie, the British published no returns of their loss, but from the vigorous resistance it must no doubt have been great. Their loss in prisoners was 385. On the part of the Americans the killed amounted to 83, the wounded to 216, and the missing to a like number.

"A few days after this battle the British raised the siege, and retreated behind the Chippewa. Meanwhile, the reinforcements from Plattsburg arrived at Sackett's Harbour, and after a few days rest proceeded to the Niagara. They crossed that river on the ninth of October, when General Izard, being the senior officer, superceded General Brown in command. On the 14th, the army moved from fort Erie, with the design of bringing the enemy to action. An attempt was made to dispute the passage of a creek at Chippewa plains, but the American artillery soon compelled the enemy to retire to their fortified camp, when attempts were repeatedly made to draw them out the following day, but without effect. A partial engagement took place on the 15th, which closed the campaign on this peninsula.

"Thus ended a third campaign in Upper Canada, without a single important conquest being secured.—The operations of the army under Brown, however, are not to be considered as worthless and inefficient. They have, in the most complete manner effaced the stain thrown on the army by the imbecile efforts of its infancy, and have cast a lustre on the American name, by a series of the most brilliant victories, over troops heretofore considered matchless. Nor ought we to lose sight of the effect produced by these events on the country at large, actively engaged as was almost every citizen, in repelling or preparing to repel, the invaders of their homes."

We have preferred to give the history of this campaign thus from a contemporary source* rather than undertake to present the same facts in different words, and perhaps, inadvertently, be guilty of inaccuracy.—This account is fortified by official documents, and may be relied upon as correct.

At the time of the battle, Mr. Gass, belonged to the first regiment, under command of Col. Nicholls; but at the time of the attack on the British batteries was attached to the gallant 21st, under Col. Miller. According to his recollection of the spot, after a lapse of forty years, a gentle slope rises within some 300 yards of the Cataract of Niagara, to the height of perhaps 200 feet, with a steep declivity at the side next the Falls.--Skirting this declivity,

---

* By an oversight discovered too late in the process of printing to be corrected, the foregoing account of the campaign on the Niagara frontier, is credited to Nile' register. It should have been to the " Historical Register of the United States," edited by T.H. Palmer, 1816. Vol.4 page 14.

there was a narrow road or rather by-path. Ascending the slope was another, more travelled road, fenced in on both sides, with a large red frame church on one side of it, at the bottom of the hill. This road was known as Lundy's Lane. After attaining the top of the ascent, the country was more level but rolling, and with patches of timber interspersed.—The British battery was located at the top of the hill, across the lane; and effectually commanding the entire ground, cut off the advance of the Americans, and left them exposed to the flank attacks of the British. This being the position of affairs, it was absolutely necessary to the Americans that the British should be driven from the position. Mr. Gass distinctly recollects of the memorable saying of Col. Miller, "I will try, sir," when ordered by Gen. Ripley on the perilous task of its capture, being of common report at the time, and voucher for its authenticity. The day had been warm and somewhat cloudy toward evening, but all was calm and summer-like,—the monotonous roar of the cataract so near at hand, drowning all meaner sounds, mingling with the thunder of the artillery, and overtopping the demoniac sounds of war. For one hour, Scott's Brigade had borne with unflinching resolution, the storm of iron hailed upon them by the battery; but it soon became apparent that the British must be dislodged or the day be lost. By this time, it was after seven in the evening,—the clouds had rolled away, and the midsummer moon nearly in its full, poured a flood of light over the battle-field. Silently and steadily the command of Col. Miller, drawn up in line but two men deep, marched up to the foot of the lane, the red church protecting there from the grape of the artillery; then, without a halt or a waver, they advanced rapidly up the hill, with bayonets at a charge, the grape flying over their heads in a harmless storm, until they gained the very muzzles of the pieces. Then, says Mr. Gass, came a blast of flame and smoke, as if from the crater of hell, and they were among the enemy,—hand to hand—bayonet to bayonet--and steel clashing on steel, in the close and murderous conflict. The fight was but for a moment. "Charge the gray backit militia:—they cannot stand the bayonet," shouted their Scottish commander, but in one moment the British were put to flight and the taunt was falsified on his very lips. Then was heard the command of the gallant Miller, "halt and form." The order was hardly executed, ere the British were back upon them like a whirlwind, and then ensued the hardest of the fight. Three several times, they made the assault and were as often repulsed. The British guns, at last were turned upon their former owners and sullenly and doggedly they were forced to retire from the field. One of the pieces of cannon, says Mr. Gass, in order to show the locality of the battery was trundled over the bank and down into the Falls. In narrating his personal experience, he says that the affair was so rapid that he hardly had time for a distinct idea, until it was over; but that in marching up to the battery, he felt as he expresses it,. "d—d bashful." We have assurance, however, that his modesty soon wore off. A ball thro' the hat, thanks to his shortness of stature, was the only mark of merit he received in this encounter. The principal carnage took place after the battery was captured; the artillery being aimed so high, as to do but little execution in the thin column of advancing Americans.—The hand to hand fight over the guns is said to have been terrific, and the bloodiest ever fought on the American continent, considering the number of men engaged and the number of the slain. The dead were literally piled in heaps. Blue uniforms and red, promiscuously mingled in the ghastly piles, and the hand palsied by death, still held the musket with its bayonet sheathed in the bosom of the foe; and the kindred blood of Briton and American mingled in one red stain upon the sodden earth. It was indeed a scene of terrible slaughter.

From 6 to 11 P. M., the battle raged about this contested spot; the placid moon looking down upon the beligerent hosts, and the stars like reproving angels, beholding the wild passions of man, thus mocking nature's thunder with his infernal din.

Pursuit was impossible, after the retreat of the British; and the American's held possession of the ground for some time; until seeing that nothing was to be accomplished by holding this now barren position, they retired in good order to Fort Erie.

The particulars of the siege of Fort Erie and the sorties made by the Americans have already been related; but an incident occurred at the memorable sortie of the 17th August, that shows the character of "Sergeant Gass" to a striking advantage. To each company was attached men whose duty it was to carry a supply of rat-tail files and a hammer with which to spike such cannon of the enemy as they should be so fortunate as to capture. Sergeant Gass, was intrusted with this responsible duty, by Capt. Denman, in whose company he served, and on one occasion having taken a small British battery, the Americans were marching off their prisoners, when Denman despatched the Sergeant to Gen. Brown, who was standing on a log, some yards from the spot to enquire whether he should destroy some 24 pounders. "Destroy them, Sergeant," said. Brown, "we don't know how long they'll be ours."—Patrick says he slapped in the rat tail files and drove them home; while some "tall yankees from York State" sledged off the trunnions from the pieces with a marvellously good will. The selection for such a duty requiring coolness and bravery, is a high compliment to the Sergeant, and shows the estimation in which he stood among his comrades and officers.

Fort Erie was occupied by the Americans until the close of the campaign, when it was destroyed and the army prepared to spend the winter at Buffalo and other points. Mr. Gass, under Gen. Winder, passed the winter at Sackett's Harbor. Having passed the winter here, without extraordinary incident, he was discharged from the service in June 1815, news of the treaty of Peace of the previous 24th December, having come to hand in February, 1815—the battle of New Orleans of the 8th January, having been fought after the treaty was actually signed by the commissioners at Ghent.

CHAPTER V

CONCLUSION—IN RETIREMENT.

The war was now closed, and our hero with many others, was thrown again upon the world, none the better for his camp service either in pocket or in morals. Discharged at Sackett's Harbor, he took up big line of march, once more, for Wellsburg. By the way of the lake, then on foot, then riding in a wagon, the old soldier returned from the wars, until he reached Pittsburgh; thence, descending the Ohio, it was not long until he again greeted his friends, to engage no more in the perils of war, nor to leave them again, in the wild search for adventure. He had taken an active part in the most arduous campaign of the war, and bad participated in its most brilliant victory; but while the laurel wreath hung so gracefully about his brow, he had also felt some of the stings of the thorn. The congratulations of his friends were embittered with the thought that now forty years of his life were passed, and he had nothing substantial to show for recompense--nothing laid up against the day when penury might plead in vain with cold-hearted charity, for alms in consideration of self-sacrifice and gallant deeds in the country's service.

He was now a middle-aged man, and very naturally began to think of making some provision for the future. Accordingly, as the phrase goes, he settled down. His subsequent career has been that of an old soldier, subsided into the realities of every-day life, and struggling against poverty for an honest subsistence. The wild oats he had sown in his early manhood, were now to be reaped. A life of settled industry was irksome to his temperament, and altogether contrary to his habits. Like too many others in his position, he give way to intemperance, and during the succeeding forty years of his life, occur many chapters, over which we gladly draw the veil of charity. We would not say a single word derogatory to temperance as a virtue, nor would we mitigate by a single iota, the proper abhorrence of vice in any shape; but surely some charity can be extended to the veteran, whose youth up to mature manhood, had been spent in the camp, and meritoriously in the service of his country. That man has in his composition little of the milk of human kindness,--little of the spirit of Him who said, writing upon the sand, "he that is innocent among you, cast the first stone," who cannot find in his heart something to extenuate, if not excuse this single failing in a character otherwise unblemished. Let us not judge too harshly. We know not the temptations of other men, nor is it ours to judge their faults or foibles too severely. Still, while we would crave a charitable judgement, the fact cannot be denied, that, like too many others, he acquired, during his campaigns, a taste for intoxicating liquors, and was, for many years, a slave to the debasing habit that degrades and demoralizes so many of the best, most brilliant and most generous of our race. Intemperance was his besetting sin, but drunk or sober, he was ever honest, sincere and truthful, and a patriot to the very core of his heart. In his very worst degradation, there was ever displayed an inherent nobility of character, which commanded the pitying respect of his acquaintances; and which in later years, has enabled him in a great measure to throw off the debasing habit.

From the time of his return in 1815, he has been located in this vicinity, engaged in various occupations. He tended ferry for 'Squire Robert Marshall, in 1815, tended Brewery for Wright & Russell, in Wellsburg, in 1816, and in the same year helped John Brown to build the old Baptist Meeting House, hunted stray horses about Mansfield, Ohio, in 1817, and labored on his father's farm, and in the fulling mill alternately, the succeeding years, until his father's death, which occurred in 1827. On this event, be was appointed administrator of the estate, which business was soon adjusted, his father's estate being but small, and Patrick's share but a trifle.

At this point commences the romantic portion of his career. He had attained to the mature age of 58, without having ever had his susceptibilities sensibly touched by the boy-god; until he was deemed impervious to his shafts, and insensible to the charms of female society. He had taken his position in the innumerable army of old bachelors, and was deemed incorrigible by his acquaintances and the gentler sex. He who had fought the wild bears of the mountains, slept with the buffalo on the plains, straddled the Missouri, and lived for months on unseasoned dog-meat, then faced the British at Lundy's Lane and Erie, and fought his way through blood and flame, it was little thought would ever surrender his manhood to weak woman's wiles and winning ways. But they misjudged him, as they were ignorant of human nature. Love goes by contraries. Like seeks not like, but seeks its opposite; so that the blended elements may make the perfect being:

> *"Breasts which case the lion's fear-proof heart. Find their loved homes in arms where tremors dwell."*

So Shakespeare says, and so in this case the event demonstrated. He had only deferred his destiny, because he had not yet found his counterpart.

In the fall of 1829, he commenced boarding with John Hamilton, better known among our younger readers as the Judge, whose bowed frame will be well remembered as he sat about the stores and street corners—a wreck of a powerful and once influential man. At this time, Hamilton lived on a piece of land, and had to cheer him a pretty daughter, whom he called Maria. She was just blooming into womanhood, and thrown into the constant society of our hero, a mutual feeling sprang up between the two, and gradually June melted into December. Of the process of their courtship we have no data other than what probability suggests. He doubtless wooed her with "tales of hair-breadth scapes, and of perils by sea and land," and as she listened, she doubtless breathed the wish, as maidens often do, "that heaven had made her such a man." Whether she did or not, they made each other understood by some subtle alchemy to lovers known; and not to theorize too far on so delicate a subject, they were married in 1831. Patrick immediately rented a house from a certain Crickett, who resided on the Crawford farm, in the vicinity of Wellsburg, and commenced house-keeping.—Maria made him a good and loyal wife, and in testimony thereof, presented him with *seven children,* during the fifteen years of their married life, from 1831 to 1846, when she died. It was customary to joke the old soldier on his rapid increase of family. Such jokes were always good naturedly received, and he would characteristically remark, that as all his life long, he had striven to do his duty, he would not neglect it now, but by industry make amends for his delay.

In his married life he was kind and affectionate-- a good husband and father. Five of his children are still living, one having died in infancy and another a well grown lad, dying in London county, Va., of the small pox, in 1855. After various changes and removes, he finally purchased a piece of hill-side land on Pierce's Run, in Brooke county, and sat down with his increasing family to cultivate the soil. This happy retirement was interrupted in 1846. At this time the measles appeared in his family—all of the children were prostrated, and in the February succeeding, came the severest blow he had ever experienced. At this time his wife having taken the measles, died, and he was left with a large family of young children dependent upon him for support in his old age.

In consideration of his services he received from the Government, in addition to his pay as a soldier, 160 acres of land in 1816, and a pension of $96 per year, to date from that period. The land he suffered to lie, until eaten up and forfeited from non-payment of taxes, and the pittance of $96 per year is all that he has actually received from the Government in exchange for the services of the best years of his life, from 1804 to 1815 over and above his pay and rations as a soldier.

The history of the pension laws of the United

States is one of interest, and notwithstanding the fact that all has not been done that gratitude perhaps demanded, she has been more liberal in this respect, than any other country in the world. It has been the rule, in all countries, to grant pensions, in some shape, for meritorious services, to acknowledge or stimulate merit, and to raise those who have served their country faithfully, above the caprices of fortune. In England, the king has been regarded as the sole judge of desert, and following out the theory of sovereignty, in America, the people have exercised the grateful prerogative. As the gratitude of the country toward the veterans of the revolution was great, so their liberality in the early history of the republic was generous beyond precedent, the more especially as the public lands furnished an apparently inexhaustible magazine of largess whence to draw. Pension acts were passed during the war of the Revolution, providing adequate support to those who might be disabled in the discharge of duty. Subsequently these laws were enlarged and explained. In 1818, those "who served in the war of the Revolution until the end thereof, or for the term of nine months, or longer, at any period of the war *on the continental establishment,*" and "by reason of reduced circumstances in life," were in need "of assistance from the country for support," were provided for. In 1828, pensions were given, without any qualification as to property, to all officers and soldiers who served in the continental line of the army to the close of the war. Finally, in 1832, the terms were enlarged, and pensions were granted to all who served in a military capacity, during the war of the Revolution, for a period not less than six months. First, those disabled in the military and naval service received pensions; then the indigent and necessitous; and lastly all were embraced.

The act of 1832, was very comprehensive in its provisions, yet in some respects it was unjust—for instance: The rate of pension was graduated by the length of service and the grade or rank in which it was rendered. Two years' service entitled the party to the full pay of his rank in the line, not to exceed, however, the pay of a captain. For shorter periods the pension was proportionably less; but no pension was provided for merely being in a battle, or for any length of service less than six months. This of course cut off a large class of soldiers equally meritorious, but whose service perhaps only extended to a single campaign or to a single battle, although that campaign of six weeks or single battle may leave been equally arduous and dangerous to the individual, as in other cases might have been the full period of the war to other individuals. Many persons, were called suddenly into active service during the war of 1812 as at New Orleans and other places, and actually engaged in active battle, perhaps been wounded and disabled, yet these men, under the provisions of the act of 1832, were entitled only to a pittance proportioned to the excess of service over six months. This, was manifestly unjust and to remedy the injustice, and in some manner equalize the public bounty, was the object of the old soldiers meeting on the 8th January, 1855, in which Mr. Gass, with many others, figured at Washington City, as hereafter narrated.

No man ever served his country more faithfully than Mr. Gass, and though humble and uncomplaining, no one ever better deserved to be a recipient of the public bounty. Had he been a titled soldier, his extraordinary claims to consideration would ere this have forced themselves upon public attention, but the most of his career was in the capacity of an humble private, without commission and without honor, save that which comes from the honest and faithful discharge of duty in whatever position he happened to be placed. Many a man of less real merit, and very many of much less service have risen higher in political and military and civil station, but it has been his to see his inferiors overtop him in the rewards as well as in the plaudits of a well spent life. He was too modest to thrust himself forward among the brazen ranks of aspirants for political preferment, too proud to crave as a largess from the government more than what he deemed himself entitled to by the terms of his contract, too much of a philosopher to complain at neglect, and too long a soldier to repine at the inconveniences of a lot which he knew to be inevitable. With the pittance of $96 a year, which he has been for many years in the habit of drawing in half yearly instalments from the agent of the government at Wheeling, and the small amount he has been enabled to eke from his spot of stony land, he has lived in patriarchal simplicity, scrupulously honest, owing no man anything, and apparently contented and happy as a millionaire. We doubt, indeed, whether ever the possessor of a fortune led a more contented or equable life. So far as worldly cares are concerned, as to himself, he lived the life of a philosopher, satisfied that he would have enough for a decent subsistence while he lived, and friends enough to give him honorable burial when he died; and hence troubled himself but little about the accumulation of property. His wants were but few, and easily supplied. But as his family crew up, and the necessity of some provision for them began to occupy his mind, it would be singular if he did not feel some degree of solicitude in their behalf. It is a beautiful characteristic of our nature, the feeling which induces us to provide for those who are to come after us, around whose lifes the chords of affection have been so entwined, that after death, we would still have them remain interlaced with the recollection of the love we bore them.—Man alone, of earthly creatures is immortal, and man alone, of all God's creatures, provides for his offspring by force of an instinct which reaches not only beyond the period of maturity, but beyond the grave. It is hard to find a creature so debased, so cold, so destitute of the ordinary feelings of humanity, as not in some degree to recognize the promptings of natural affection and in many a one, the secret of a long life of toil, of trouble, of peril and deprivation, of sacrifice of comfort and even of character, is found in this yearning after posthumous regard. Undefined, the feeling may be, perverted and wrong it very often is, yet such is the case--that around the most rugged heart, the desire for providing competence for posterity, has grown into a passion, until it has choked out almost all other kindly feelings, and the man becomes a miser, that his children may be spendthrifts. The feature, in human nature, of which this is an exaggeration, in its true and proper developement, beautifies and ennobles man and distinguished him from the brutes that perish. Mr. Gass, had now become a man of family, and as the cares of providing for them began to thicken around him, he began to be more solicitous for a proper provision for their welfare.

During the year 1854, the propriety of calling a convention of the surviving soldiers of the war of 1812, to meet at Washington City, by delegates, to memorialize

Congress for some further acknowledgement in the shape of grants of lands, of the services of those who had served the country in its day of adversity, was actively discussed. The country had now grown strong and wealthy, and it was thought that of the present abundance we could well spare some for the relief of the surviving and in many cases fortune broken soldiers. The case certainly appealed strongly to the generosity, if not to the justice of the country. A strong feeling seemed to exist on the part of the public to recognize these claims. The question was discussed in Congress, and advocated on the ground of sound policy as well as of gratitude, and the Press of the country was nearly unanimous in favor of the propriety of acknowledging the claims of the old soldiers. President Pierce in his annual message of this year spoke of the commendable policy of setting apart a portion of the public domain for this purpose and adverted to the fact that since 1790, 30,000,000 acres of public lands had been appropriated for the benefit of those who had served in the wars of the Revolution. Nothwithstanding this disposition, however, there were still many to object to such a measure. It was urged that the pension laws were liberal already beyond those of any other country; that a further extension of liberality, would open the door to corruption and fraud; that many of the surviving soldiers were wealthy and did not need the bounty, and that it would be hard to discriminate between them who did and those who did not; that it would be unjust to favor same merely because they were *survivors*, while others equally meritorious, had died without such favor; and that to equalize the matter it would be necessary to extend the bounty to the heirs of the latter, which would absorb too large a portion of the public lands. These were serious objections and prevented congress acting as liberally toward the old soldiers as otherwise their feelings might have prompted them to do. Nevertheless, it was thought advisable for these latter to visit Washington City in person and thus appeal to the country for an extension of liberality in their behalf. Accordingly, public and very general notice was given by advertisement and circulars that a convention of old soldiers would assemble at Washington City on the 8th. day of January, 1855, and the surviving soldiers were invited to assemble in their respective neighborhoods and send on delegates to represent them at this general meeting. This call emanated from the president of the "military convention" of the soldiers of 1812, which had assembled in Philadelphia the 9th. of January preceding and contemplated a meeting not only of representatives of the surviving soldiers, but of the heirs of the deceased soldiers. It was particularly desired however, that as many of the old soldiers as could make it convenient should attend in person and by their presence make the demonstration the more impressive and effective. In accordance with this call, a meeting of the old soldiers of the vicinity was invited to meet at Wellsburg on the 25th. of December 1854, to elect delegates to the National convention of the 8th. January. The names of those present at this meeting are as follows, as they stand in the published proceedings of the meeting in the Wellsburg Herald of that date.

Patrick Gass, Maj. John Miller, William Tarr, Isaiah Roberts, Robert Britt, Walter Brownlee, Eli Green, Obed Green, Wm. Roberts, Noah Barkus, George Young, Mathias Ebberts, Ellis C. Jones, Elijah Cornelius, Wm. Cole, John Moren, James Davis, James Wells, Walter D.

Blair, Adam Ralston, Wm. Atkinson, James Baird, John Gatwood, and eight others were represented by their nearest male relations. The scene presented by the assemblage of these gray haired veterans, some of them trembling at the very verge of the grave and none of them with more than a very few years of this world in prospect meeting thus in council, was unique and suggestive of the times that tried men's souls. Conspicuous among them appeared our old friend Mr. Gass, to whom was assigned the post of honor at the head of the list in consideration of his eminent services as well as of his age. The tall form of Ellis C. Jones towered among his old comrades—venerable with his snow white locks and the casual observer could readily recognise in the faces of many of them the evidence of that manly energy that prompted them in their youth to stand in the fore front of battle in defence of the country. Though old, crippled by disease and time and accident, there was still about them something that distinguished them from the mass of their fellow men.

The meeting appointed Messrs P. Gass, John Miller, Wm. Tarr and Ellis C. Jones to represent them in the old soldiers meeting of the 8th. January ensuing.

The Convention met at the 4 1/2 Street Presbyterian Church in Washington City, on the morning of the 8th, and organized by the election of Joel Y. Sutherland, of Philadelphia, as President, when after prayer by Rev. Sunderland of Washington, they were addressed by Peter Wilson of Cayuga, and others, on the subject for which they had assembled.

After adjourning, they formed in procession and preceded by all the military of the city, and various bands of music, marched to the President's house, which they reached at about 2 o'clock, and found the President and most of the Cabinet in waiting to receive them. President Pierce, in response to the address of the President of the Convention, delivered an appropriate speech, substantially as follows:

"I tender to you, sir, and to your associates, my grateful acknowledgements for the privilege of this interview, and for the kind reference you have been pleased to make to myself. It can hardly be necessary for me to say that my heart sincerely responds to your allusion to the hero, who has given immortality to the day, you have met to celebrate. As this numerous assemblage of veterans filed before us, no man could have observed their countenances, without being impressed with the fact that they were the men for such a war as that of 1812. The lines of intelligence and marked emphasis of character are unmistakeable. What a crowd of associations spring from the presence of the veteran commanders near me, (General Scott and Commodore Morris,) and I am gratified to observe among you a delegation from our red brethren, who were found faithful in the period of trial; and whose services are entitled to be cherished with grateful remembrance. Many of you have never met before, since the close of the war, and this reunion of companions in arms must revive in your bosoms, gentlemen, emotions peculiarly active. My earliest reading was of the occurrences of forty years ago, in which you all bore a part, and my earliest reminiscences are of the war of 1812. I well recollect that approach of every mail was anticipated by my footsteps to the village post office, and that I naturally felt the deepest concern for those who left my own home to take a part in the conflict, while my young heart gave out its quick sympathies to all who contributed to the cause, personal service, or sustained it at home by earnest and efficient encouragement. Time has only served to enhance the admiration I then felt, for such as promptly enrolled themselves under the flag of their country, and it is gratifying to meet here to-day, so many survivors of that gallant army and navy. I can readily conceive the thrilling emotions that must rush upon you as you now grasp each other by the hands, for the first time for forty years, and it may be for the last time forever; but, gentlemen, I must not detain you. I wish for you, individually and collectively, every blessing—all that you can reasonably expect, and all that your country can consistently confer. The universal commendation which greets you at every step, is more eloquent than any words that I can utter. May God, who has so signally blessed our country, preserve and ever bless its defenders."

Six cheers were then given for the President, and as many more for the soldiers of 1812, and the convention, after the war-chief of the Onondagua had replied to the President, adjourned for dinner.

At 5 o'clock, the convention having resumed its deliberations, passed the following preamble and resolutions:

"Forty years have now elapsed, since General Jackson fought and won the last great battle of the late war with Great Britain, at New Orleans; and that glorious anniversary is a fitting day for the surviving soldiers of that war to meet and take counsel together. This city, too, named after the father of our country, is a most appropriate place for our assemblage. The war of the revolution achieved our liberty, the war of 1812 secured it. While the green sod marks the graves of our revolutionary fathers, a few only of those who staked their lives in our last immortal conflict, survive to tell the tale of our sufferings and services—by far the greater portion of them having passed down to their last homes on earth, many of them in penury and want.

"In less than forty years after the close of our revolutionary struggle, a grateful Congress passed a general pension law for the benefit of the surviving officers and soldiers, at a time when the treasury was empty, and a heavy war debt was hanging over it; is it then unreasonable for us to expect that similar justice will be done to the survivors of the war of 1812, and to the widows and children of those who are dead, while the public treasury is overflowing with gold and we have comparatively no debts? Or is it asking too much to have fair portions of the public domain, which we fought and paid for, allotted to us? We think not.

"1. Be it therefore, resolved, that a committee be appointed to memorialize Congress on this subject, and to urge upon our Senators and Representatives to make to each officer, soldier, sailor and marine, who served during the war of 1812-15, appropriate grants of land,—least 160 acres to the lowest grade and for the shortest time of actual service. The benefit of this law to extend to the widows and children of those who are dead.

"2. That similar provision ought to be made for our red brethren who fought by our sides; and all those confined in foreign prisons during the war of 1812, if alive; and also, to the prisoners in Tripoli, who were forced to labor as slaves or felons. If dead, then to their widows and children.

" 3. That while we deeply deplore the untimely deaths of so many of our brethren in arms, we pledge ourselves ever to aid and protect their bereaved widows and orphans; and here on this most interesting occasion, we extend to each other the right hand of fellowship, and bind ourselves by every sacred obligation to stand by each other while we live, in defence of all our rights at home and abroad.

"4. That Congress ought to extend to the soldiers of the late war and their widows the same pension system, adopted for these of the revolution; and the thanks of this Convention are hereby tendered to those just and generous members of both houses, who have had the nerve already to move in this matter.

"5. That in our judgement every principle of justice requires that *invalid* pensions should commence from the time when the wounds were received or disabilities incurred in the service of the United States."

Resolutions were also passed thanking individual members and recommending measures to be taken to carry out the objects of the foregoing resolutions, after which speeches were delivered by Generals Scott and Coombs of Kentucky, and by other distinguished soldiers and civilians, when the Convention adjourned until the next morning.

Next morning, the old soldiers again assembled, and after prayer proceeded to discuss the most available means for accomplishing the object of the Convention.

Committees were appointed for every State, and arrangements made for an organized effort to secure tardy justice from the representatives of the people. After some time spent in discussion they adjourned, having been kindly treated by the citizens of Washington, and being highly pleased with their entertainment generally.

The parting of the old soldiers when each delegation took up its line of march for home, was affecting in the extreme, and as the gray haired veterans shook hands for the last time many an eye was bedewed with tears. Since that last parting many of them have died, and as year by year rolls around one by one the defenders drop off, until now their ranks are more than decimated—scarcely enough remaining to call the roll of the survivors.

The meeting of the old soldiers was a failure so far as the object for which it was previously designed, was concerned; but it attracted the attention of the country to the subject, and may yet eventuate in the desired modification of the pension laws, or at least to an extension of liberality to particular individuals like our hero, the singularity of whose service precludes the probability of his case ever being used as a precedent. It would be creditable to the American Congress did they make an exception in his case, and by a bonus, munificent to him, trifling to the nation, demonstrate at once their appreciation of meritorious services and character, and help to smooth the declining years of an old and deserving soldier.

There is not probably now living, a single man who has done so much for the public as Mr. Gass, and received as little. Among the many unique features of his character this is not the least singular. He has never been a beggar, neither has he ever had emolument thrust upon him by the country he so faithfully served; hence he is both poor and humble. It may be proper, now, to say, that these suggestions are ours, not his; not put forth at his instance, but unsolicited, and wholly prompted by a desire on our part to see a proper liberality extended to a deserving man. As for him, his desires are but few as his wants are simple; and if the government begrudges a material acknowledgement of his claim, we are satisfied that the refusal will not occasion him a single pang of regret, or a single murmur of complaint.

After his return from the "Soldier's Convention" of 1856, although disappointed in his anticipations, he manifested a philosophic indifference, and much more surprise and gratification at the development of the country, the magnificent railroads, public buildings and improvements that came under his observation, and the universal recognition that he received among the dignitaries at Washington City, and indeed among all parties with whom he came in contact, than he did regret or dissatisfaction at the result. During his travels about the Federal city he was considerably lionized, had the freedom of the various routes of travel, and generally was highly flattered by the consideration that was shown him on all occasions where his character was known. A considerate friend in the city had presented him with a spread eagle of brass which was attached to the front of his hat and wherever the badge was observed it became his generally recognised passport.—The same badge was worn for a long time after at home, and regarded by the old hero, with very commendable pride as a souvenir of the kindness of his Washington friends.

We shall now bring the biography of our hero to a close, only remarking that as the blemishes on his character are few and superficial; the reader who sincerely appreciates his really good and sterling qualities, will readily forget and forgive his frailties.

    He is still living, December, 1858, a hale, hearty Virginia Democrat of the old school,—one who never faltering in the discharge of duty, or deviating by the breadth of a hair from the strict line of principle, still mingles suavity with his party zeal; and that grateful tolerance of opinion in others, which distinguishes the gentleman from the mere politician. He is one of nature's gentlemen, is the least that can be said of him. Having nursed James Buchanan in 1794, he of course, voted for him for President in 1856, as he has uniformly done for his democratic predecessors since the days of Andrew Jackson. His political views are firm and decided, but he seldom obtrudes them; his religious convictions are of the same cast—immovable, but undemonstrative. Such as he is, he stands before the world: and such as he has been, he is prepared to go before his Maker in full reliance upon his justice and grace, without meddling much with creeds or professions. So, we leave him. That his latter days may be prosperous and happy, and blessed with the christian's hope of immortality, is the sincere wish of his biographer.

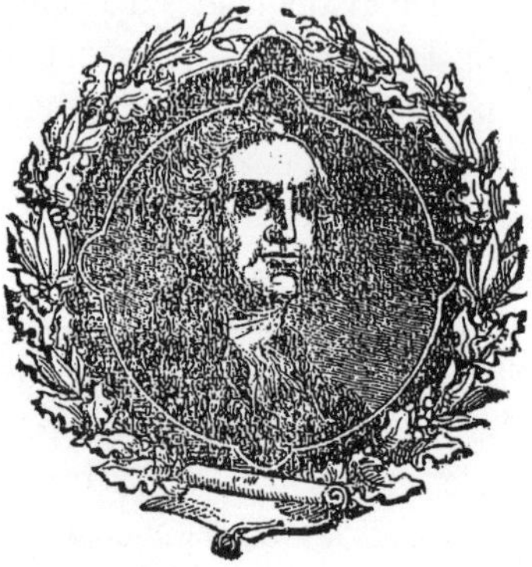

PART SECOND.

CIVIL HISTORY.

The Upper Ohio—French and English Pretensions—Washington's First Expedition—First English Settlement at Fort Pitt—Governor Dinwiddie—Washington's Second Expedition—Fort Necessity—Death of Jumonville—First Gun of the 70 year's war—Washington's Capitulation at Fort Necessity.

It may not be inappropriate to the design of our work, to give some detail of the civil history of the country coming within its scope, the more particularly as writers, heretofore, have generally contented themselves with the more heroic features of our annals as exemplified in the narratives of Indian wars and massacres.—This is a much easier style of composition than the laborious collation of facts and figures and as a general thing a more interesting one to the cursory reader, who by the way, represents the large majority of the reading public. The civil history of the country, nevertheless, is important; and may be made interesting. The materials in the crude state may be found pretty widely diffused through the public archives, in the columns of old newspapers, in private repositories of papers, and in the memories of contemporaries. To all these sources we have resorted when opportunity offered, and one result of our researches has been, a knowledge of the wide difference between the random stringing out of words and correct statement of facts capable of being verified by comparison with dates and authorities. We have been able to discover no regular history of this character, and believe the items have never been systematically collated. Few sections of the country can boast of more incident in its early settlement than that lying on the upper waters of the Ohio, and it may be necessary to preface our civil history with some of rather a martial east. It was the theatre of controversy between the French and the English prior to the Revolutionary war and even before it attracted English attention, was regarded with covetous eyes by the French government. They contemplated a chain of posts extending from the lakes to the Gulf of Mexico by means of which they might be enabled to gain and preserve the supremacy of the country. Their object was principally trade with the Indians, though political reasons and perhaps religious proselytism, were impelling motives for their actions. The point of confluence of the Allegheny and Monongahela was early and rightly considered a most eligible situation for a stronghold commanding as it did, the mouths of two rivers along whose banks the peace and war paths of the Indians of the North and West concentered, and being at the head of the most magnificent water course in the world, 3000 miles in length, and then considered much longer. It was rightly considered the key of the western country. Both the French and the English saw its importance and both were disposed to take measures to secure possession of it. As early as 1753-4, Washington at the age of 21, had been sent by Governor Dinwiddie, of Virginia, to inspect it. He pronounced warmly in its favor as an eligible place for a military post, and recommended its immediate possession. He also gave it as his opinion, that the point would some day be the seat of a great city. In May, 1752, the Indians, by treaty at Logstown had "desired their brothers of Virginia to build a strong house at the forks of the Monongahela;" and at Winchester in 1753 another party had renewed to Virginia, the same proposal. They were afraid of and angry at the French; and courting favor with their competitors, the English.

The Ohio company, in the early part of 1753 had opened a road from Will's Creek into the valley of the Ohio, and in November of this year, the young Envoy, with Christopher Gist as guide, an interpreter, John Davidson by name, and four attendants on horseback and on foot, travelled in nine days to the forks of the Ohio. The season was cheerless, with sleet and snow and the prospect gloomy with the fallen leaves and the solemn silence of the late Autumn, but the prophetic mind of Washington grasping the future, was able to overlook the inconveniences and drawbacks of the present, in the magnificent country that opened upon his vision along the banks of the beautiful river. Pursuing his journey, he held favorable council with the Indians at Logstown and Venango, but was able to effect nothing with the French, whose commander St. Pierre, an officer of courage and ability, bluntly informed him that "he was there by the orders of his General to which he would conform with exactness and resolution, and that he would sieze every Englishman within the valley of the Ohio." One object of Washington's embassy was to ascertain the object of the French in encroaching upon the territory in time of "solid peace" and their answer was satisfactory upon that head. This took place at Fort Le Boeuff, or Waterford, 15 miles south of Lake Erie, on French Creek, and immediately retracing his steps he started about the middle of inclement December, back for Virginia. The cold increased very fast and the wilderness paths were obliterated by the deep snows, so that they were compelled to travel by compass alone. The day after christmas, while travelling he was aimed at by an Indian at fifteen steps distance, but the gun missed fire; then they started across the Allegheny on a raft of logs, constructed with infinite trouble, with the aid of "one poor hatchet," and when in the middle of the running ice, Washington was jerked overboard by catching his setting-pole between two large cakes, and saved himself from drowning only by grasping the logs of the raft, and lodging upon an island. The next morning, the Allegheny was frozen and they finished the perilous ferriage over the ice. By January 1754, they reached Gist's settlement at the foot of Laurel mountain, and after that, their progress to the seat of government at Williamsburg was less arduous.

His report was followed by immediate activity, even on his return he met pack horses laden with materials and stores and families going out to settle at the Forks of the Ohio, as it was at that day called. The Ohio company had somewhat anticipated his report. They commenced the Fort and made some progress when Contrecoeur came down from Venango, with field pieces and near 1000 men in sixty bateaux and 300 canoes, and demanded its surrender. Having only 33 effective men, they, on the 17th of April, 1754, capitulated and withdrew. Contrecoeur finished the fortifications and named it Fort Duquesne. In the meantime, Gov. Dinwiddie had been exerting himself to forward soldiers to the scene of operation. Capt. Trent was commissioned to proceed forthwith, and having raised a company of 100 men, ordered to march to the Fork, and complete the Fort, and Washington was authorized and directed to recruit a force at Alexandria, for the same purpose. But difficulties occurred in the colonial Government, Capt. Trent proved inefficient, and before efficient aid could be rendered, the fort had fallen into the hands of the French. It was the first regular English settlement on the waters of the Ohio. Gov. Dinwiddie was disposed to take vigorous measures for the settlement of the country. Two hundred thousand acres of land lying on the Ohio river, one hundred thousand lying contiguous to the Fort for the use of the garrison, were offered as an inducement to volunteers. This proclamation was effective, and is the foundation of the titles of many of the farms lying in this region. Two dollars per hundred acres was afterwards the price fixed by the government for warrants for unappropriated lands, located in any quantity and almost anywhere. The offer of bounty induced ready enlistment, and on the 2nd of April, 15 days before the fall of the Fort, Washington set off for the forks of the Ohio, with 150 men, and was followed by Col. Fry with the remainder of a regiment. They experienced great difficulty, had to impress horses and wagons and got bad ones, the roads were miserably bad, and on the 9th of May they were still nine miles distant from Will's creek fort, at a place called the Little Meadows. By the 27th they had descended the waters of the Youghiogheny, until they came into close quarters with the French. Warned by the Half-king, a friendly Indian, whose friendship Washington had gained in his previous excursion to the Ohio, and by his old friend, Christopher Gist, near whose residence he then was, to be on the alert, he halted at the Great Meadows and proceeded to fortify his position. He named the place Fort Necessity. The French were under the command of de Jumonville, a young officer of great promise. Washington and his party, assisted by the Indians under the Half-king, surprised them in their encampment and after a short encounter, in which ten Frenchmen were slain, and twenty-one taken prisoners, defeated them.—Jumonville was killed at the beginning of the skirmish, and his death was made the theme for much declamation. The French court denounced the act as contrary to all the laws of war and claimed that Jumonville and his party were only engaged in a peaceable embassy and were on the search for Washington and his party, whom they had heard of, as being on the way. This was afterwards proved to be all pretence, and of a piece with the dissimulation which the French habitually practiced in their proceedings, during this controversy.

Washington, himself, fired the first gun, and says Bancroft, "his word of command kindled the world into flames. It was the signal for the great war of the Revolution. There in the Western forest began the battle which was to banish from the soil and neighborhood of our republic the institutions of the middle ages, and to inflict on them fatal wounds throughout the continent of Europe. In repelling France from the basin of the Ohio, Washington broke the repose of mankind, and awakened a struggle which could admit only of a truce, until the ancient bulwarks of catholic legitimacy were, thrown down."

It may thus indeed be said that on the waters of

the Ohio, was the first gun fired of the war of opinion that afterwards convulsed the world, and whose reverberations did not cease until the American colonies were freed not only from French but from English dominion; and Europe itself was shaken to its centre by the armed hosts of seventy years of almost continuous war. Through all its vicissitudes, the conflict of liberty with legitimacy, was the prevailing idea—culminating in the attrocities of the French revolution, and expiring from exhaustion alone, with the fall of the first Napoleon. The death of Jumonville was hailed all over both continents as the first overt act of hostility between France and England, and commenced the "old French war," which resulted in the supremacy of England in the valley of the Mississippi, and proved the nursery for the gallant soldiery who in after years in turn, wrested its possession from her, and in the name of the people, took charge of it themselves.

After this affair at the Great Meadows, Washington determined to push on toward the Forks, and proceeded some distance, but ascertaining that the French would meet him with an overwhelming force, he judged it best to retreat, which he did, until he again reached Fort Necessity on his return. His men were jaded and discouraged, and scarce of provisions, and he waited for reinforcements from Wills creek with supplies.—While waiting at this point, a deserter carried word to the French under de Villiers, a brother-in-law of the de Jumonville previously slain, of the desperate condition of the Virginians, and that officer at once marched to attack them, and on the 3d of July, after a severe conflict, Washington capitulated, obtaining favorable terms from the French commander. On the 4th, they took up their line of march again from the valley of the Ohio, as prisoners of war; and the French flag waved undisputed by any actual force from the head springs of the Ohio to the mouth of the Mississippi.

  A copy of the articles of capitulation was subsequently laid before the Virginia House of Burgesses, and notwithstanding the unfavorable termination of the enterprise, Washington and his troops were thanked for their gallant behaviour and about $1100 (300 pistoles) voted to be distributed among the men engaged.

CHAPTER. II.

THE INDIANS AND THEIR, POLITY.

Policy of the Indians—The Ohio Indians—Hunting Grounds—Shawanees—Ottowas—Six nations—Indian Villages—Tecumpseh—Hatred of the Whites—Cruelties—Pioneers—Bounties for Scalps—Indian war 1768 Comparative Losses—Scouts.

History does not speak in very favorable terms of the conduct of the Indian allies of the English. Notwithstanding all their exertions and the expenditure of a large sum in presents to the Indians, not more than thirty could ever be obtained, at one time, to join the forces of the English in this campaign. They appear to have been regular mercenaries, easily discouraged by adversity, and difficult to control in time of success, apt to desert when most needed, and generally willing to sell themselves to the highest and best bidder. The English and French bid for their services. The former had early gained the good will of the Six Nations, as they were called, by timely assistance afforded them against their enemies, the Adirondacks, who were aided by the French; while the latter, by their superior diplomacy and greater versatility of character, gained aver the good will of the Ottowas and Northwestern Indians with whom they traded and trapped and intermarried. The French could always turn their Indian allies to better account than could the English; and on several occasions had large numbers of them in service, and used them to great advantage. Contrecoeur's successful expedition against Fort Pitt, is a case in point, as is also Braddock's defeat, and the engagement with Col. Boquet, in which the French and Indians were, however, defeated. In each of these affairs, the Indians greatly outnumbered the French. At this time, the French had also alienated several tribes of the Six Nations from their old friends, the English, though they were unable to retain them until the end of the war.

As the Indians played an important part in the early settlement of this section, and the details of their wars with the whites, compose a good portion of our early history, we shall indulge in a few remarks and reflections upon the circumstances of their existence in the land. The country lying on the waters of the upper Ohio does not appear to have been very strictly appropriated by any particular tribe of Indians, but to have been regarded as a common hunting-ground for all.—The mountainous and hilly region of the Monongahela and Allegheny, with its numerous streams, abounding with game and fish, was roamed over by parties of all the tribes for a great distance around. The numerous stone arrow-heads turned up in every new-ploughed hill-side and top throughout this wide region, is evidence that it was industriously hunted, while tradition reports that at an early day, the creeks and rivers literarily swarmed with fish of the finest kinds. The conformation of the country rendered also its valleys and ridges the thoroughfares for Indian parties travelling from one section of the country to the other, on their various excursions and they had well beaten paths in every direction. Tumuli and mounds exist in abundance, and along the river bottoms the disinterment of Indian remains are of frequent occurrence. All this goes to show that the country was much frequented by the Indians, still it does not appear to have been the seat of any considerable villages during the memory of the whites, at least. Small settlements of a few huts like that at Logstown, Catfish Camp, the Mingo bottom, and others existed, but rather at the head quarters of some noted chief or warrior, than as the settled habitation of any tribe. It may be, that the continual liability of the country to be over-run with hunting parties, often of hostile tribes, prevented it being more densely populated, certain it is, that the principal Indian villages whence came the savage irruptions into the infant settlements of Virginia, and Pennsylvania were situated far distant from this locality. Of all the Indian tribes of which we read, the most unrelenting, and apparently the most numerous and powerful appears to have been the Shawaneess who dwelt upon the Miamis and the flat lands of Central Ohio, extending to the Wabash. They were originally Tennessee Indians, driven thence into the Ohio country by the Creeks at a period not very remote. Next to them in importance, appear to have been the Delawares, a powerful tribe driven from the Susquehanna country by the encroaching whites and located near neighbors of the Shawanees with whom they acted often in concert. Next, the Ottawa's, a large and enterprising tribe inhabiting the lake country to the Northwest and the land of the Illinois, and after them the Wyandotts, Mingoes and a score of others—smaller tribes—some of whom had their villages on the banks of the Ohio and tributary streams, but who were not generally considered very formidable. To the Northeast, lay the country of the Six Nations, along the lakes, the St. Lawrence and the Hudson, capable of bringing into the field 2500 fighting men. The Mohawks were the most noted among those confederated tribes. Among people so nomadic in their habits as these, it is difficult to assign any very definite boundaries, but they appear to have had certain rules and regulations among themselves which were scrupulously observed. Each tribe appears to have had a certain territory and villages peculiar to itself, where the families, the old men and the infirm resided and to which the warriors and hunters repaired as to a general rendezvous, but the country outside of this appears to have been occupied and hunted in common. This idea of community of lands, seems to have been a prevalent one in Indian polity. Particular tribes had particular local habitations over which they claimed and exercised exclusive jurisdiction; but all the balance of the land was a common, to which all the individuals of all the tribes in the country among whom peace prevailed, had certain common and undisputed rights, which could not be violated without offence. We find Tecumpseh, the great Shawnee chief, who was dissatisfied with the treaty made between the Indians and Gen. Wayne, after their disastrous defeat in 1792, urging as the cause of his dissatisfaction that the tribes who were parties to the treaty, disposed of privileges to which they had no right. At the Council of Vincennes he claimed for all the Indians of the country a common right to all the lands in it; denied the right of any tribe to sell any portion of it without the consent of all; and therefore, pronounced the treaty of Fort Wayne, null and void.

Such was the state of this section at the time of which we speak in reference to the Indians. It was overrun by wandering bands of Indians of divers tribes and language, often at war with one another and not very scrupulous upon whom they committed depredations, but particularly jealous of the whites, whom they all regarded as intruders upon their common territory.—It was difficult to effect treaties, and when violated, it was extremely difficult to ascertain and punish the violation. They were in regard to the whites, more like irresponsible banditti than anything else. Distrustful alike of the French and English, and hating equally both; they were willing to lend themselves to whichever paid the best or promised most opportunity for taking pale-face scalps. The Indian, naturally blood-thirsty, had in this case, both tradition and his own knowledge to encourage him to hate the whites. They would willingly have exterminated them, but they are cunning and crafty, as well as brave; and revengeful, and easily appreciating the hopelessness of open hostility, they were disposed to effect the same end by stratagem and management. It was a pleasure for them to see their white brothers engaged in throat cutting, as it saved them the trouble and the risk. They fought on the side both of the French and the English. At Fort Necessity, Washington was aided by the Indians, and at Braddock's defeat but a few months afterward, the same Indians assisted the French in the slaughter of the English. Said the Half King, the "French were cowards and the English fools." Crafty, bloodthirsty and cruel, yet endowed with many virtues, among which were desperate courage and tenacious patriotism, they were enemies not to be despised, and friends, whose alliance was to be courted and purchased. With all their faults it cannot be said of them that they were regardless of the faith of treaties when properly understood and fairly treated; or that their cruelty in war was unprovoked. The whites in both respects have much to answer for. The Indians were the original possessors of the soil, and the whites could advance no stronger claim than they, hence they were disposed to regard with jealous alarm the pretensions of the English and French to the possession of all their territory, even from the rising to the setting sun. Their jealousy to say the least, was but natural, and much allowance is to be made for them in their pertinacious assertion of what they deemed their rights. Still, the Indians had no equitable title to all the territory over which they saw fit to assert a claim, any more than had the whites; and it is a narrow-minded philanthropy that regrets their being dispossessed of a land they could neither appreciate or improve. The Indian, in the wisdom of Providence, had fulfilled his destiny; a stronger and a subtler race from beyond the great waters, had come to push him toward the setting sun, and though he might struggle and writhe in his savage agony, yet the advancing wave in its irresistable majesty swept him before it, or mercilessly buried him and his, with the memory of his ancestors in the gulf of oblivion. Had they done otherwise than they did, they would have been more or less than men; hence we are disposed to look leniently upon Indian barbarities, and with a philosophic eye upon the causes and the manner of their extermination.—Divested of romance and poetry—the two races were antagonistic in almost every respect—they could not exist in peace together—and the weaker yielded,—is the philosophy of Indian history, condensed.

But however philosophic in theory, they were solemn realities to the pioneers. As early as the day of which we write, scattered families of whites driven by the love of adventure, or fear of justice, or allured by fabulous accounts of the fecundity of western soil, had located upon the banks of the Ohio, the Monongahela, the Allegheny, the Kanawha, the Holston, the Potomac and their tributary streams. Distant, hundreds of miles from each other, they were liable to be murdered in their cabins by the marauding Indians, and their fates perhaps never be known. Such cases, there is reason to believe did happen.

During the peace preceding the French and Indian war just inaugurated, these settlements had increased in number, but when the war removed from the Indians, what little restraint they were previously under, settlements ceased, and Indian murders became so frequent that the country was nearly, if not altogether abandoned by these sentinels of Civilization, and the clearings left to grow up with weeds and underbrush, to be again reclaimed in happier times. Bounties for scalps were offered by both parties, to their disgrace be it said, and the vindictive Indian took a savage delight in the silken locks of women and children, as well as in the scalps of his more legitimate victims.

From the commencement of hostilities, the country was a continued scene of warfare in detail; but in 1763, the Indian war may be said to have commenced in earnest, when the Indians significantly left a tomahawk in the cabin of a murdered family near fort Ligonier, as a formal declaration of war. Shawanees, Delawares,

Mohawks, Wyandotts, and Mingoes, all seemed to unite in a war of extirpation. The whites, says Col. James Smith, of Kentucky, a veracious man, and for many years an adopted captive among the red men, lost in the ratio of ten to one. Lurking parties attacked them in their cabins; they skulked around the homestead and shot the farmers at their work or while hunting or journeying; they waylaid the emigrant by water, and as the descending craft swept with the current against the projecting headland, it was assailed with rifle bullets by unseen enemies. Under such tuition, our early settlers became almost Indians themselves in their watchfulness and keen sagacity as scouts, as well as in unrelenting hatred of their enemies.

It is no part of our plan however, to depict the horrors of Indian warfare. Others have given in detail the barbarities of both sides; for ourselves, we would willingly they were forgotten, for there is little in the record of attrocities to elevate our conceptions of human nature or to improve or elevate the race. In many cases cold blooded butcheries were perpetrated by the whites, and in some cases, without the shadow of a justification other than passion and revenge. Such enormities were incident to the war, and we turn from them in preference to the more grateful record of the peaceful progress of the country.

CHAPTER III.

BRADDOCK'S EXPEDITION.

Confidence of the French—Stobo—Gen. Braddock—Sir John St Clair—Provincials disgusted—Departure of troops—The "Black Rifle" Difficulties of the route—Battle Ground—Battle—Fall of Braddock—Washington to the rescue—Braddock's Death—French and Indians—General Panic—Pontiac's War—Emigration stopped—Col. Boquet's Stratagem—French Supremacy Wm. Pitt—Gen. Forbes—Fort Duquesne retaken—Fort Pitt.

The subsequent warlike proceeding of the English and Virginians in their efforts to dispossess the French from the valley of the Ohio, are so intimately connected with our early history, that we feel constrained to trace them further. After the defeat of the Virginians at the Great Meadows, and expulsion from the country, the French appear to have relaxed in their vigilance at Fort Duquesne, so that Stobo, one of the two hostages left as security for the fulfillment of the articles of capitulation, found means to send to the government at Williamsburg, a map of the fortifications and a detailed statement of the strength and disposition of the garrison. Induced by these representations, it was determined to make an effort to retake it from Contrecoeur.

The colonial government, although Governor Dinwiddie found it very impracticable as a general thing, voted 20,000 pounds sterling and the home government furnished about the same amount in money and arms, for the purpose of carrying out the design. Major General Edward Braddock, a veteran of forty years standing in the most precise school of British discipline and exact punctilio, was appointed to conduct the campaign. Braddock was brave and kind hearted, an experienced soldier according to routine; but obstinate, overbearing and lacking in common sense to appreciate the difference between war in civilized style and war in the wilderness. Sir John St Clair, deputy quarter master of the expedition and a man of much influence in it, was an obstreperous, swearing Briton of pretty much the same character as his superior. These two officers moulded the character of the campaign. A regulation of the Government degrading the colonial officers below officers of the same rank in the King's regiments had so disgusted Washington, that he had retired from the service. When the British fleet however, with two prime regiments of well equipped soldiers landed in the Chesapeake and the brilliantly equipped soldiery were disembarked at Alexandria, almost within sight of his home at Mt. Vernon, it so stirred the military ardor of his blood that he was readily induced to listen to overtures flattering to his pride; and to accept a place in Braddock's staff. A convocation of the governors of the different colonies met at Alexandria to concert measures for the campaign. The result of their deliberations on the point in question, was, that Braddock set out from Alexandria, on the 20th April 1755, in great state attended by a military cavalcade for the rendezvous of the forces at Wills

Creek. By the 30th, of May, after much delay and embarassment the troops were all at Wills Creek, ready for their march, to the number of nearly three thousand men, of whom, about one half were British regulars.

They had not proceeded far on their route before the General discovered the nature of the enterprise in which he had engaged. The Provincial officers would come to him with advice in his dilemmas but with a strange perversity, he spurned their counsel as presumptuous and insulted some of them by imputations of caution amounting to cowardice. Captain Jack, a bold and intrepid borderer, known in early times as the "Black Rifle" and a terror to the hostile Indians, tendered his services and was rebuffed by the over-confident general and turned on his heel with his band of a hundred leather clad rangers and disappeared in the woods.—He would have been of the greatest assistance, had he continued with the troops. As his difficulties increased Braddock condescended to consult with Washington. At his suggestion, twelve hundred men of the choice of the army were told off, to march as rapidly as possible toward the Forks, while Col. Dunbar was left behind with the balance to make the road and bring on the heavy artillery and baggage. This plan promised success. The army advanced much more rapidly; the expedition having consumed nearly a month in accomplishing one hundred miles. At length, on the 8th July, they had reached within fifteen miles of their destination. Scouts had been out constantly, and Christopher Gist returned in the morning from the immediate vicinity of the fort, narrowly escaping with his life from a couple of Indians, and reported the road clear and no enemy to be seen in force about the premises. At the point they then occupied, the hills came down bluff to the water, forming a narrow pass of some two miles on the side of the river on which the fort was situated, which it was considered dangerous to attempt; and it was resolved to ford it and march down some five miles on the other side and again recross. At day break, the next morning, the troops were put in motion. They forded the Monongahela with all the precision and deliberate confidence of a parade. Their arms glittered in the sunlight and their accoutrements were all in faultless order, as they formed on the opposite bank and marched along the open valley. The officers were all in full uniform and all looked as if arrayed rather for a fete than for a battle. Washington, who had been sick and left behind to recover, at Fort Necessity, and had rejoined them but the day before still indisposed; smarting under the contemptuous rejection by Braddock of his cautious suggestion that he should keep the Virginia rangers in advance of the regulars, as more accustomed to the mode of warfare, nevertheless looked upon the pageant with an admiring eye. Roused to new life, he forgot his repulse and all his recent ailments and broke forth in expressions of enjoyment and admiration as he rode in company with his fellow aids de camp, Orme and Morris. Often, in after life, he used to speak of the effect upon him, of a well disciplined European army, marching in high confidence and bright array, on the eve of a battle.

About noon they reached the second ford, Gage, with the advance, was on the opposite side of the Monongahela, posted according to orders; but the riverbank had not been sufficiently sloped. The artillery and baggage wagons, drew up along the beach and halted until one, when the second crossing took place, drums beating, fifes playing, and colors flying as before. When all had passed, there was again a halt close by a small stream called Frazer's Run, until the General arranged the order of march.

First went the advance, under Gage, preceded by the engineers and guards, and six light horseman, then Sir John St. Clair, and the working party with their wagons and two six pounders and on each side were thrown out four flanking parties. Then at some distance, the General was to follow with the main body, the artillery and baggage preceded and flanked by light horse and squads of infantry; while the Virginia and provincial troops, were to form the rear guard.

The ground before them was level until about half a mile from the river, when a rising ground covered with long grass, low bushes and scattered trees, sloped gently up to a range of hills. The whole country, generally speaking, was a forest, with no clear opening but the road, which was about twelve feet wide, and flanked by two ravines concealed by trees and thickets. It was now near two o'clock. The advance party and the working party had crossed the plain and were ascending the rising ground. Braddock was about to follow with the main body and had given the word to march, when he heard a quick and excessively heavy firing in front. Washington, who was with the General, surmised that the evil be had apprehended had come to pass. For want of scouting parties ahead, the advance parties were suddenly and warmly attacked. The firing continued with a fearful yelling. There was a terrible uproar. The general sent forward an aid to ascertain and report to him the cause, and too impatient to wait spurred after his messenger. The turmoil increased. The van of the advance had been taken by surprise. It was composed of two companies of carpenters or pioneers to cut the road and two flank companies of grenadiers to protect them. Suddenly the engineer who preceded them gave the alarm, "French and Indians." A body of these latter was approaching rapidly, cheered on by a Frenchman in a gaily fringed hunting shirt, who was slain in the charge and proved to be the commander of the attacking party, Captain de Beaujeu.

There was sharp firing on both sides at first and several of the enemy fell; but soon a murderous fire broke out from the ravine on the right of the road, and the woods resounded with unearthly whoops and yellings. The Indian rifle was at work, leveled by unseen hands. The advance was killed or driven in. Gage ordered his grenadiers to fix bayonets and charge up a hill on the right whence there was the severest firing. Not a platoon would move. They were dismayed and stupified as much by the yells as by the rifles of the unseen savages. The latter extended themselves along the hill and in the ravines; but their whereabouts was only known by their demoniac cries and the puffs of smoke from their rifles. As the covert fire grew more intense, the trepidation of the regulars increased. They fired at random whenever they saw a motion and shot some of their own flanking parties and of the rangers who had like the Indians, taken to the trees and were doing good execution. All orders were unheeded. The officers were doubly exposed and in a very short time were most of them shot down. The advance fell back upon Sir John St. Clair's corps, which was equally dismayed.

Col. Burton, had come up with the reinforcements, and was forming his men to face the rising ground on the right when both of the advanced detachments fell back upon him, and all now was confusion.

The Virginia troops, accustomed to the Indian mode of fighting, scattered themselves, and took posts behind trees where they could pick off the lurking foe.— In this way they in some degree protected the regulars. Washington advised the General to adopt the same mode with the regulars, but he persisted in forming them into platoons; consequently they were cut down from behind logs and trees as fast as they could advance. It was little better than murder for men to be thus exposed. Some of them attempted to take to the trees without orders, but the general stormed at them, called them cowards and even struck them with his sword.

The slaughter among the officers was tremendous.—They behaved with the most consumate bravery. In the desperate hope of inspiriting the men they could no longer command, they would dash forward singly or in groups. They were invariably shot down; for the Indians aimed from their coverts at every one on horseback or who seemed to have command. Some were killed by their own men, who crowded in masses, fired with affrighted rapidity. Soldiers in the front were killed by those in the rear. Between friend and foe, the slaughter of officers and men was terrible. All this time, the woods resounded with the unearthly yelling of the savages, and now and then, one of them, hideously painted, and ruffling with feathered crest, would rush forth to scalp an officer who had fallen, or sieze a horse galloping wildly without a rider. Such is a description of the battle as depicted by the graphic pen of Irving. Such an unmitigated slaughter could not long continue. Nearly all the regular officers were disabled, the troops were paralyzed by the panic, all subordination was lost, Braddock with obstinate bravery still attempted to retrieve the fortunes of the day, when a bullet, aimed, it is doubtful whether by friend or foe, passed through his right arm and into his lungs, and he fell from his horse, having already had five horses shot under him. In his despair he wished to be left upon the field to die, but was with difficulty removed. The principal command now devolved upon Washington. Throughout the day, he had signalized himself by his calm courage and great presence of mind. He exposed himself without reserve to the murderous rifle, and his escape seems little short of miraculous. Two horses were killed under him, and four bullets passed through his coat, nevertheless, he escaped unhurt.

After the fall of Braddock, the rout was complete. Baggage, stores, artillery, everything was abandoned. The wagoners, took each a horse out of his team and fled. The officers were swept along in the headlong flight.—The Indians rushed from their coverts, and pursued the frightened fugitives, as they clashed across the river, in the tumultuous confusion, killing many while in the stream. A body of them were rallied at a spot about a quarter of a mile beyond the river, where Braddock had been conveyed, and an effort made to effect a stand, small parties were told off, and sentinels posted, but before an hour had elapsed, most of the men, sentinels and all, had stolen off. Being thus deserted, there was no alternative, but a precipitate retreat.

Washington was sent back to Dunbar's camp,

forty miles distant, to carry the news, and to Hurry forward provisions, hospital stores and wagons for the wounded, but the tidings had reached Dunbar before his arrival, and the camp was wrought into the greatest trepidation by the exaggerated reports of the frightened fugitives, and it was with the greatest difficulty a precipitate flight was prevented by the officers.

The detachment escorting the wounded General, augmented to a couple of hundred men and officers, reached Dunbar's camp, on the 12th, and on the 13th, the entire force took up its melancholy march, back again to the Great Meadows, which they reached in the evening. Here, Braddock died, on the night of the 13th.—His proud spirit was broken by defeat, and the difficulty with him seemed to be to comprehend how it came to pass. He was grateful for the attentions paid to him by Captain Stewart of the Provincials and Washington, and more than once it is said, expressed his admiration of the gallantry displayed by the Virginians in the action. It is said, moreover, that in his last moments, he apologized to Washington for the petulance with which he had rejected his advice, and bequeathed to him his favorite charger and his faithful servant, Bishop, who had helped to convey him from the field. His obsequies were performed in sadness and before the break of day, Washington reading the funeral service in the absence of the chaplain, who had been wounded, and his grave was carefully concealed so as to avoid desecration by the lurking savages. The place of his sepulchre, however, is still known and pointed out and we believe a monument now marks the spot. The last acts of his life displayed the real generosity and kindheartedness of the man; while his dauntless conduct on the field, shows him to have been a fearless as well as an accomplished soldier. Technical pedantry and military conceit were the chief errors of his character, and bitterly he expiated them by an unhonored grave in a strange land, a memory clouded by misfortune and a name forever coupled with defeat.

Out of eighty-six officers seventy-two were killed or wounded, and of the rank and file upward of seven hundred. The Virginia corps suffered terribly. One company was annihilated, another had but one officer left and he, a corporal. Their assailants were but a comparative handful, being not the main force of the French, but a detachment of 92 regulars, 146 Canadians and 637 Indians, 875 in all, led by Capt. de Beaujeu. Contrecoeur, had received information that the English 3000 strong, were within eighteen miles of his fort. Despairring of making an effectual defence against such a superior force, he was balancing in his mind whether to destroy the works and retreat, or to stay and obtain honorable terms. In this dilemma, Beaujeu prevailed upon him to allow him to sally forth with a detachment to form an ambush and give check to the enemy. His request was granted as a sort of forlorn hope. Beaujeu, not having time to complete his ambush, the attack was precipitated and Beaujeu fell, almost at the first fire. His Indians, however, spread themselves among the trees and logs along the whole length of the army and in a marvellously short time the whole line was at once assailed. Their rifle shots spread like wild fire, and the woods resounded with their yells. Then, ensued the panic and slaughter—as the Indians expressed it, they shot them down "same as one pigeon."—The whole number of killed and wounded of the French, and Indians, did not exceed seventy. No one was more surprised than Contrecoeur himself, when the ambuscading party returned in triumph, with a long train of packhorses laden with booty, the savages uncouthly clad in the garments of the slain—grenadier caps, officers gold laced coats and glittering epaulets, flourishing swords and sabres, or firing off muskets and uttering fiendish yells of victory. But when he was informed of the utter defeat of the British army, his joy was complete, he ordered the guns of the Fort to be fired in triumph, and sent out troops in pursuit of the fugitives.

We have thus been particular in narrating the fate of the Expedition;—in doing which we have drawn largely upon Irving's Life of Washington—because, in the first place it is a notable incident in our history, and in the next, because it was followed with most important consequences to the country at large. Braddock's defeat elated the Indians, and encouraged them to carry desolation even beyond the mountains. Winchester was threatened, and the valley of Virginia was almost deserted of its inhabitants, emigration of course stopped and the prospect was gloomy in the extreme. If it was not the severest check British power ever received on the continent, it was certainly the most humiliating. The entire campaign was a compound of mismanagement, cowardice and misfortune. Its consequences ran forward into the revolution. The militia ascertained that the British regulars were not by any means invincible, and gained confidence in themselves and in their officers, while the attempt of the home government to compel the colonies to pay part of the expenses of this very expedition, was a prominent incentive to the rebellion of 1775.

From this time until 1763, raged what was called

Pontiac's war, one of the most awful periods of distress ever before or afterwards experienced in the western country. This was closed by the decisive victory of Col. Boquet at Brushy Run in Westmoreland county, Pa., in the August of that year, which so dismayed the savages that they gave up not only all further designs against Fort Pitt, and the surrounding country, but withdrew temporarily from the frontiers. In this engagement, the Indians were themselves ambushed and defeated, in a style similar to that which eight years before they had so effectually used against Braddock. The English army consisting of about five hundred men, the remnant of two regiments of Highlanders, more than decimated by disease in the West India service and sent into the northern woods to recuperate, was marching with a large convoy of stores, through the wilderness on the 4th of August 1763, with no appearance of an enemy in sight, when suddenly at midday, the advance as in Braddock's case, was violently attacked. But the Highlanders, better prepared than their predecessors, charged them with fixed bayonets, and drove the savages before them, but with considerable loss to themselves. They fell back, and the savages swarmed around them, confident of victory, thirsting for their blood, and yelling with fiendish delight at the prospect of another Saturnalia of carnage. But Boquet was cooler and shrewder than Braddock. Perceiving the overwrought audacity of the savages, he took advantage of it. Posting two strong companies, concealed in the underbrush, at each side of his road, he commenced a precipitate and apparently disorderly retreat. The savages fell into the snare. Thinking that the English were really in confusion and retreating, they dashed yelling from their coverts, in full pursuit; when the two concealed companies assailed the exposed mass with a heavy fire on either flank; and at the signal, the retreating troops faced about and poured into the astonished Indians such close and galling vollies, that they were stricken with panic, and yielding to the irresistable impulse, were utterly routed and put to flight. It was a deathblow to the Indians and a dear victory to the English. Boquet, lost in killed and wounded, about one fourth of his men; and was hardly able to convey his wounded—stores and everything else being destroyed— to Fort Pitt which he reached four days after the battle.

From Boquet's victory, dates the undisputed possession of the Ohio valley to the white man. The power of Pontiac, the "Colossal chief of the Northwest," was broken; his adherents were dispirited by defeat, and sued of the whites for peace; but the name of the chieftain still hovers over the Northwest, as that of the hero who devised and conducted their great but unavailing struggle with destiny, for the independence of their race. In this war, they scalped over a hundred traders in the woods, they murdered many families in their habitations, they besieged and took by force or stratagem numerous forts, and slew their inmates; they threatened the very strongholds of the whites; passed the mountains, and spread death and terror even to Bedford, Winchester, and Fort Cumberland. Nearly five hundred families from the frontiers of Maryland and Virginia fled to Winchester, unable to find even so much as a hovel to shelter them from the weather, bare of every comfort and forced to lie scattered in the woods.

In the mean time the defeat of Braddock and its terrible consequences, had vacated nearly every English cabin in the valley of the Ohio. Of the North American continent of twenty-five parts, France claimed twenty; leaving but four to England and one to Spain. She had in the execution of her plan, connected the great valleys of the St. Lawrence and the Mississippi, by three well known routes,—by way of Lake Erie and Waterford to Fort Duquesne, by way of the Maumee to Shawneetown at the mouth of the Wabash, and by way of Chicago, down the Illinois, and she seemed prepared and able by arms and art, to make good her claim of possession.

The war started in America had embroiled the parent countries. Misfortune and mismanagement seemed to attend every motion of the English. They were effectually humiliated, yet with true British doggedness they were neither conquered or discouraged from further attempts at retrieving their fortunes. The ministry determined to regain and hold the supremacy of the western world. They found the colonies in their assemblies impracticable, headstrong as themselves; the spirit of independence sturdily asserting itself at every show of arbitrary power on the part of the crown. Still they persevered. William Pitt, the great commoner, who had now risen through difficulty and all manner of opposition to the ministry of England, trusted and loved by the people for his manly qualities; feared and respected by the nobility for his ability and boldness, had become the ruling spirit at home. In colonial matters, when entreated to interpose, he regarded the bickering between the people and the assertors of prerogative, with calm impartiality and blamed both parties for the failure of the English arms and policy in America. He determined to retake fort Duquesne, as a part of his far reaching plans of re-conquest. The colonies themselves felt their honor at stake, and the Assemblies seconded his determination with unusual zeal. Twelve hundred and fifty Highlanders arrived from South Carolina and rendezvoused at Fort Cumberland. Pennsylvania added twenty seven hundred men, and the "Old Dominion" nineteen hundred more, besides a corps of three hundred and fifty Royal American volunteers. This overwhelming force for the service, was put in motion under the command of Brigadier General Joseph Forbes, called the "Iron Head" an able officer but in the last stages of a fatal disease. Here, the fortunes of Washington again mingle with those of the Ohio country. He was stationed at Fort Cumberland with the Virginia troops and insisted upon advancing promptly along Braddock's road; but was provoked at the dilatory policy of Forbes, in having a new road cut through the wilderness nearly parallel therewith.

Intelligence having come to hand that Fort Duquesne was defended by but five hundred French and three hundred Indians, Major Grant with 800 highlanders, and some Virginians, was detached by Col. Boquet, without the knowledge of Forbes, to surprise and take the Fort. The vainglory of the Major, led to his defeat, and the rout of his army with a loss of 300 men; the balance being saved only by the good conduct of the provincials. Washington was then permitted to proceed with his Brigade of Provincials to attempt the capture of the Fort, the garrison of which having been reinforced by four hundred men from the Illinois, was now near twelve hundred strong. As Washington and his Brigade advancing by forced marches, and followed by the main army approached the Forks, the Indians deserted them, and on the 25th November 1758, reduced to 500 men, the garrison disheartened by the prospect, set the fort on fire, and by the light of the conflagration descended the Ohio. This took place in sight of the youthful American hero, and ere the smouldering flames of the fortress had expired, he planted the British flag on its deserted ruins. Thus, without the firing of a hostile gun, or the spilling of a single drop of blood in battle, was accomplished by the Provincial Major, and his Virginia brigade, what the martinets of the British army, with the power of England at their back, had expended hundreds of lives to accomplish, and failed in the effort. Gen. Forbes about this time, died. Thus fell French supremacy in the valley of the Ohio.

As the banners of England floated over the Ohio, the place was with one voice called Pittsburgh. It is the most enduring trophy of the glory of Wm. Pitt. "Long as the Monongahela and the Allegheny shall flow," says Bancroft, "to form the Ohio, long as the English tongue shall be the language of Freedom in the boundless valleys which their waters traverse, his name shall stand inscribed on the gateway of the west."

CHAPTER IV.

SETTLEMENTS, TITLES AND BOUNDARIES.

Early Boundary Disputes—First Settlements—Pennsylvania and Virginia State Line—Patents—Litigation—Titles—Lord Dunmore—Conolly—Revolution—Capt John Neville—Early Patriotism—Settlement of Boundary Disputes—The Panhandle—Origin of the name—Ohio county—West Liberty—Original Settlers—Characteristics—Early Enterprise—Imigration—Weighty Characters.

At a very early day, as far back, at least, as the commencement of the 18th century, disputes arose as to the title of the land lying on the waters of the Ohio, which were never definitely and authoritatively settled until after the war of the revolution. The French claimed the entire country from the mouth of the Mississippi to the head springs of the Ohio, by virtue of discovery, under the name of Louisiana, while the English claimed from Plymouth and Jamestown, west, to the other ocean, under titles claimed by the "divine right" of King James and his successors. Subordinate to these original claims were the claims of the proprietaries of the different States indefinitely worded, and of necessity, often clashing.—Some of these, again, recognised a sort of title in the Indians, which in some cases, they purchased for considerations more or less valuable, and in others, siezed, by virtue of conquest. It is believed that no white man trod the shores of the Ohio or its upper branches, prior to 1700; as early, however, as 1715-20, an occasional trader ventured beyond the mountains, and among the first of these, says the historian of Western Pennsylvania, was James L. Fort, who took up his residence at what is now Carlisle, in 1720. A Mr. Frazer was a prominent trader among the Indians, at about this date, and resided at the mouth of Turtle Creek, on the Monongahela. The settlement by the Ohio Company, previously referred to, at the Forks, may be considered as the first regularly attempted white settlement. At about this period, the entire region was generally believed to belong to Virginia—though, the grant to the proprietary of Pennsylvania, expressly guaranteed to him the country from a certain point on the Delaware river, the starting point of the celebrated "Mason & Dixon's line," five degrees of longitude west. The ideas of geography in those days, were, however, somewhat indefinite; and Virginia had counter-claims, which she put in; and at any rate, she exercised jurisdiction over all that portion of what is now Pennsylvania, included between the Monongahela and the Ohio, and an indefinite territory besides, beyond her present boundary. This entire scope of country was called West Augusta, by the Virginians, and embraced from the Blue Ridge west to the Mississippi. By a law, passed in 1769, forming the new county of Botetourt from Augusta; it being considered that the people living on the waters of the Mississippi, would be seriously incommoded, by reason of remoteness from the Court-house of Botetourt, they were considerately exempted from the payment of levies imposed for the building of the Court-house and jail. The county of Fincastle carved out of this, in 1772, was again subdivided in 1776, into Kentucky, Washington and Montgomery counties. Thus, vague and indeterminate, were the boundaries of this region, only eighty years ago.—After Fort Pitt came into the hands of the English, by the treaty of Grenville, in 1765, and during the lull in Indian hostilities subsequent to the events before narrated, emigration having again commenced, and settlements having been gradually made along the various streams, as the population increased, boundaries became of more importance. The Western portion of the district, comprising the territory lying upon and between the waters of the Monongahela and the Ohio, took the name of Yo-ho-gania, as appears by the Virginia patents of that date, which name was retained up to as late as 1785. Still, however, boundaries remained undetermined, and had become the source of frequent litigation, so that it became indispensibly necessary to settle them authoritatively, at least, between the different States.—Forty miles of territory was in the anomalous condition of belonging to two jurisdictions; the inhabitants recognizing either or neither, as suited their present inclinations. Virginia had two Court-houses South of the Monongahela, and one North, at Redstone, now Brownsville. She at one time fixed a seat of Justice at "Razortown," two miles West of what is now Washington, and at one time, the Pennsylvania proprietary offered to compromise, by making the Monongahela and the Ohio the boundary, thus taking in 'Yo-ho-gania," into Virginia. But Virginia claimed to the Laurel Mountain. The location of land warrants was the immediate subject of litigation. The Virginia laws, on this subject, were very liberal—the Indian title was now considered to have been conquered in the war, and all that was necessary to give validity to title, were such restrictions as were necessary to prevent confusion. Six months' time was to intervene between the registry of the claim at the land-office, and the issue of a patent. The patent, cost surveying and officer's fees and $2 per hundred acres of land. These conditions complied with, the patent was issued. Priority of claim was also established by "tomahawk right,"—the claimant of a particular piece of land, marking out a line through the woods by "blazing," or "chipping out," the trees around it, and deadening a few trees near a spring; and this title, although it had no legal force, was yet respected by the settlers, and became of the same force as law, as it was not deemed creditable or safe to interfere with a claim thus established. These claims were often bought and sold. The Pennsylvania proprietary, in pursuance of the policy of Wm. Penn, in 1768, went through the form of purchasing the Indian title to the same territory, instead of taking it, as did the Virginians, by right of conquest; and fixed the price of warrants under his authority much higher, the authorities say from $25 to $30 per hundred acres, or fourteen fold. He opened a land office at about this time, but the difference in price, determined the majority of the settlers to purchase from Virginia. Settlements made or warrants located previous to this date under authority of either province were recognised by both as good and valid. June 1774, a vexatious contest commenced between Pennsylvania and Virginia in relation to these matters. Lord Dunmore, was then Governor of the latter state, and as the revolution was in its incipient state and the governor a strong tory in principle and subsequent practice; it has been surmised, his object was to embroil the states in difficulties between themselves, and thus withdraw their attention from the engrossing questions of the revolution. It was at this time, that the Pennsylvanians offered to make the Monongahela the boundary line; but Governor Dunmore, evidently did not wish the dispute settled. He appointed to the command of Fort Pitt, Col. Wm. Conolly, a rash, headstrong, unscrupulous man, who harassed the people by his exactions to the point of exasperation; and even arrested and imprisoned magistrates acting under authority of Pennsylvania in the discharge of their duty. So threatening an appearance had the affair at this period that it promised to end in a civil war, and attracting the attention of patriotic citizens of both states, on the 25th. of July, 1775, the delegates in congress, including among others, Thomas Jefferson, Patrick Henry, and Benjamin Franklin, united in a circular, urging the people in the disputed region to mutual forbearance. Says the circular: "We recommend it to you that all bodies of armed men, kept up by either party, be dismissed; and that all those on either side, or in confinement, or on bail, for taking part in the contest, be discharged." To such a pitch did the mutual acrimony of feeling reach, and so disagreeable was the continued disputation that about this time, it was seriously contemplated by many of the settlers, to move in a body farther west; and a scheme with this end in view was actually entered into by a Mr. Jackson, which however failed.

When the revolution actually broke out, the attention of the people was turned in that direction. Dunmore developed his character in espousing openly, the cause of the crown against the people; and in attempting to incite the negroes in one section, and the savage in another, against the whites. At this juncture it became necessary to forget their bickering and unite for mutual defence against their common enemies--the British, Tories and Indians.

Mutual danger and a common cause united them, when appeals to their reason and patriotism were unavailing. Virginia, still claimed and exercised jurisdiction throughout the revolution, and sent out Captain John Neville with a small military force, to occupy and hold Fort Pitt. He appears to have been a prudent and conciliatory man; at any rate, the difficulties appear to have been greatly modified under his administration. It seems to have become gradually understood on both sides that it was wiser for then to defer until a more auspicious period the settlement of the boundary line; and to unite with all their zeal and energies in the common cause. It is an acknowledged fact that the cause of the revolution had no stronger friends, than among the settlers of western Pennsylvania and Virginia.—They were whigs by birth and education, and though their Irish blood made them contentious in time of peace they were united as one man against their hereditary oppressor in time of war. Says the eloquent historian. "We shall find the first voice publicly raised in America to dissolve all connection with Great Britain, came, not from the Puritans of New England or the Dutch of New York, or the Planters of Virginia but from the Scotch Irish Presbyterians," such as peopled the vallies of the Ohio and its tributaries, at this day. Under the kinder feelings produced by united resistance to Great Britain, movements were made toward the close of the war to effect an amicable settlement. For this purpose, George Bryan, the Rev. Dr. John Ewing, and David Rittenhouse on the part of Pennsylvania, and Dr. James Madison, late

Bishop of the Protestant Episcopal Church in Virginia, and Robert Andrews, on the part of Virginia, were appointed, in 1779, commissioners to agree upon a boundary. They met at Baltimore, on the 31st of August, 1779; and in 1780, entered upon their business, by continuing, according to agreement, concurred in by the Legislatures of both States, "Mason and Dixon's line," five degrees of longitude, west from the Delaware river, thence to the northern boundary of Pennsylvania, to constitute the boundaries of that State. But pending operations, the surveyors were compelled to suspend, owing to the hostility of the Shawnees and other Indian tribes, who, considering themselves overreached by the whites, in a treaty of that year, threatened to kill any surveyors, whom they might find in the territory, and consequently, continued their northern survey only to its point of intersection with the Ohio, at the extreme end of what is now Hancock county, Va. Their report was received, and ratified by the Legislature of Virginia, on the 8th of October, 1785, and from that day, dates the legal existence of the 'Panhandle." Previous to this, Ohio county had been formed from Yo-ho-gania, by the line of Cross Creek, and says the record, on the settlement of the boundary question, in 1789, that portion of Yo-ho-gania, lying north of this creek, was added to Ohio, being too small for a separate county, and the county of Yo-ho-gania became, thereupon, extinct. Hancock, then, and so much of Brooke as lies north of Cross Creek, was the last of the ancient Yo-ho-gania. Tradition, in accounting for the strip of land, driven in wedge-like, between Ohio and Pennsylvania, constituting what is called the Panhandle, states that it was, owing to an error in reckoning, that the five degree of west longitude, reached so far to the west, and that much dissatisfaction was excited, when the result was definitely ascertained. Great importance was undoubtedly attached to the command of the Ohio river, by the authorities of either State, but it is doubtful whether the Virginians felt themselves, at that juncture, very far overreached. It will be borne in mind, that at that day, the Northwestern Territory, comprising the great State of Ohio, was an integral part of the Old Dominion, so that even under the arrangement agreed to by the commissioners, the Ohio, for a very great distance, flowed through her territory, and it was not until the cession of that Territory, in 1789, some years after, that she realized the hard bargain, thus unwittingly made. When the State of Ohio was established in 1802, the Panhandle first showed its beautiful proportions on the map of the United States. A long and bitter dispute was at any rate, happily settled by mutual concession, to be only casually disturbed during the railroad era of 1854 by a slight movement toward annexation to Pennsylvania in consequence of alledged legislative neglect and grievances. It gave, what perhaps few of the people interested, expected, not only Pittsburg and its environs and all Allegheny and Westmoreland counties; but all Washington, Fayette and Green, to Pennsylvania. The Virginians, in the event, undoubtedly had the hardest of the bargain, though at the time, they did not forsee the result, or anticipate so much liberality in their future legislation.

After the boundary question became satisfactorily settled, the small strip of land running up between the Pennsylvania line and the Ohio, settled up more rapidly than any other portion of Northwestern Virginia. Having had the public eye directed to it by the many disputes, it attracted the more attention, and figures extensively, at an early day, in Legislative annals.--From its peculiar shape on the map, it received the name, in Legislative debate, of the Panhandle, given it by John M'Millen, delegate from Brooke, to match the Accomac projection, which he dubbed the Spoonhandle. The Virginians, were a little sore when they made the discovery that they were over reached by the Pennsylvanians in the bargain; but were consoled by George Mason with the reflection that the narrow strip left them, would serve the purpose of a sentinel and protect the body of the State from any invasion of its territory or institutions. However it may be as regards the interests of the State, the connection of the Panhandle country in its detached condition has not been of any material advantage to the section itself, but has rendered it liable to all the odium among citizens of the free States that attaches to slavery; and at the salve time, has rendered it impracticable for its inhabitants to avail themselves of any of the advantages of that institution. Not only that, but being so isolated, it has little in common with the balance of the State; and its inhabitants cannot reasonably expect to receive a proportionate share of advantage from the system of public improvements for which the State has made such lavish expenditures. Nevertheless, unless it has been of very late years, the people of the Panhandle have not been behind any of their fellow citizens in regard and attachment to the institutions and laws of the old Dominion; nor have they ever shown any deficiency in the article of State pride, that so pre-eminently distinguishes the Virginian, wherever and however he may be located.

On the first development of the Panhandle, it

constituted a portion of the extensive county of Ohio, which dates back to before the revolution, and reached territorially to an indefinite extent. On the waters of Short Creek, celebrated from the earliest period for the exceeding richness of the soil, was located the seat of justice for this immense territory. It was called West Liberty, and here on the 16th January 1776, was held the first Court for Ohio county, and perhaps the first civil Court ever held in the valley of the Mississippi. A court house and jail were ordered in the following spring, and among the attornies practising, are the names of Philip Pendleton and George Brent in 1778. The town was incorporated November 29th, 1786. At the organization of the present county of Brooke in 1797, at which period the seat of justice for Ohio county, was removed to Wheeling, and at about which time the county records were burned, West Liberty, was quite a metropolis, and was the scene of many a hard fought battle with forensic as well as physical weapons. The court-house, or the relics of it may yet be seen, being a log building; nearly opposite the tavern stand known as " Bill Irvin's." In its precincts, Doddridge, Sprigg, Fitzhugh, M'Kennan, and many another, whose name has since become classic, thundered their eloquence, and plead for justice and their clients. The venerable spot is also associated in the minds of the older men of this day, with many a rough joke, and row, and drinking bout. It was a great place for horse-racing, and the present generation of its citizens, came honestly by their critical love for this noblest of animals. Nor were militia musters the mere scoff of boys and ridicule of men; but something substantial. The fuss and feathers of military parade sat much more appropriately upon men who had drawn sword in the revolution, and tracked the wild Indian, with rifle cocked, ready to tree and fire, at the rustling of a leaf, than upon the holiday soldiers of to-day. The pioneers were given too, we are sorry to say it, to their grog. Not such vile compounds of strychnine, tobacco, and alcohol, as their descendants too much affect; but pure old rye, honestly distilled, by men who were as honest as their grain, and too unsophisticated to be guilty of rascally adulteration, even had they had the villainous components. Still, they drank too much,—albeit, their whiskey was good. It has been observed that although men drank freely in those days, and were frequently drunk; yet, when they became sober, no evil effects followed the potation,—the toper recovering at once, his wits, and his vigor of body and mind, instead of being shattered and besotted in nerve and intellect. It is certain that the pioneers enjoyed, many of them, rugged and uninterrupted good health, to the end of very long lives. We have listened to the recitals of the deeds of the notables of that day, until we seemed translated back to the good old days, when jollity and good neighborhood, and generous deeds, tempered the rudeness of our fathers, and men seem actually to have lived more for sociability, and for one another, than for themselves and money. In the sterling qualities of a manly character, they certainly excelled their descendants. It is true, that they had their vices in those days; but they were the vices peculiar to a new country, and to an unorganized state of society. The turbulence and lawlessness that sometimes prevailed at their gatherings, is not surprising, when we recollect that it was no unusual thing for two thousand men to assemble about the court-house at West Liberty or on occasion of a general muster; and in such a mass of semi-wild characters, gathered from the woods and hills and hollows for many miles around, it would be singular if no outlaws could be found. For them, however, justice was both sharp and quick. If we are to believe tradition, forty fights a day was no unusual thing on such occasions; the performances occasionally varied with a free fight, in which the crowd participated *ad libitim*. Their fighting, however, seems to have been more an innocent way they had of working off their surplus pugnacity than an exhibition of the ugly element of malice that generally gives point to such exercises in our day. The point of honor was settled by a passage at arms after the most approved style of backwoods etiquette; and when once decently decided, the parties shook hands, took a rousing drink all round, and from that time forward were considered as good friends as though nothing had occurred between them.

The state of society generally, in this section eighty years ago, was very similar to that which now prevails upon the outskirts of our newly settled states; with perhaps the exception of containing a larger infusion of the fighting element than in these latter, owing to the almost continual conflicts of the settlers, first with the French and Indians, then with the Indians, and finally with the British during the war of the revolution; for it must be borne in mind, that the men of whom we treat, were the cotemporaries of Morgan, Campbell and Lewis, of King's Mountain, and Point Pleasant; and many of them, held commissions under the sign manual of Washington himself, or had borne arms in the "brave old continentals."

The old settlers of this section were largely Marylanders, Virginians, and North Carolinians; and naturally introduced into their new settlements, the manners and customs of the hospitable and never over industrious sections whence they came. Many of them first saw the country during the French war, when they were induced to enlist in Col. Fry's regiment at Alexandria, by the promise of land about the Forks of the Ohio; others were induced to emigrate by the Ohio Company; and others, again, came voluntarily, because it was a goodly land. Those who came under the provisions of Dinwiddie's offer of land, secured their warrants and after the termination of the Indian wars, proceeded to locate them. Washington, himself located largely in Western Virginia from having his attention directed to the country during his earlier services, prior to and during this French war. South of Marshall county or the base of the Panhandle, the country however, was slow about filling up—population tending more toward the north.

The different settlements appear to have been made by people from neighboring localities, the ties of friendship and kindred, with apprehensions of danger, inclining them to set their stakes in close communities. A squad of Marylanders would settle here, a company of Virginians there, while in another section we would have an Irish settlement, and in still another, a detachment of Germans or Scotch; and to this day, these localities are distinctly marked by peculiarities of names, manners and modes of speech. The Short Creek country about West Libercy, early attracted settlement by its fabulous fertility, and was appropriated by horse-racing, fox-hunting, jolly Marylanders and Virginians—some of them, men of much education and refinement, and early given to hospitality, good living, fun and intermarriage. Farther north, the Scotch and Irish element began to predominate, though the prevailing type, continued Virginian. Among the original settlers of Ohio county, may be named Jas. Caldwell, George McCulloch, Benj. Biggs, And. Woods, John Boggs, Joseph Tomlinson, Ebenezer Zane, Moses Chapline, John McColloch, Solomon Hedges, John Williamson, David Shepherd, Archibald Woods, Z. Sprigg, Alexander Mitchell, &c., whose names appear prominently on the record; while in 1787, several patents were located in Brooke, or Yohogania, by Dorsey Pentecost, Moses Decker, Peter Cox, Benjamin Wells, John Van Metre, Benj. Johnson Jr., who was a surveyor, and located 7000 acres in 1785, Wm. McMahon, who appropriated the hills lying back of Wellsburg, in 1786, Hezekiah Hyatt, Lawrence Van Buskirk, John Buck, and Gabriel Greathouse, besides many others, whose names do not figure so prominently. These appear to have constituted the advance guard of pioneers, for after their arrival, there was a cessation of entrys, until 1795, when it again commenced in redoubled numbers. Among this latter irruption we find prominently the names of Thos. Cook, Nathaniel Fleming, Jas. Darrah, Wm. McClane, Benjamin Reed, and others. An esteemed correspondent in this connection, furnishes the following reminiscences:

"Among the pioneer citizens who made their first western location in the border village, we have heard the names of Col. McKennon, father of the late Hon. T. M. T. McKennon, of Washington county, Pa., who died at Reading, Pa., July 9th, 1852, universally respected and regretted. Judge Alexander Caldwell, subsequently of Wheeling, Va., Col. Oliver Brown, a distinguished officer of the Revolution, and a member of the Boston Tea-party. Rev. Dr. Joseph Doddridge and his brother Philip Doddridge,

Esq., Alex. Wells, the father and patron of Methodism in Wellsburg, and his son Bazaleel, then a young lawyer from Maryland, Charles Hammond, Esq., late of Cincinnati; Colonel Thorpe, Nicholas P. Tillinghast, Zaccheus Biggs, and many others equally respectable and influential. In the vicinity of the place, lived Capt. McMahon, who lost his life while serving in the army of Gen. Wayne, the Cox's, the Swearingen's, the Brady's and others, whose names are intimately associated with Border history.

"In the neighborhood lived, lang syne, some families whose hospitalities were so courteously and liberally dispensed, as to tempt the elite of the town to make frequent excursionary visits among them; for there, in addition to the attraction of social intercourse, they enjoyed pure air, green fields, and substantial fare; no trifling considerations to persons confined to the treadmill recreations of a small village. Upon one occasion it was the good fortune of our humble self to make one of such a party, the tableau of which is now vividly present to our memory. It was to the manor of Geo. Hammond, a Virginia Magistrate, and father of the late Chas. Hammond, Esq., of Cincinnati. Mr. Hammond was a Marylander, and a true gentleman of the old school type. His intelligent and expressive eye, silvery locks, tall, erect figure, cane in hand—inspired the beholder with feelings of reverence and veneration, while listening to his judicious and instructive conversation. He seemed to be perfect master of his establishment, which, in addition to a large family of sons and daughters, included quite a number of well fed, glossy-faced Africans."

Under the operation of the very liberal Virginia laws regulating claims to unappropriated lands, the good land of the country was rapidly taken up, and generally in large bodies, by the parties named above, and their cotemporaries--a large proportion of it on speculation, to be sold at an advance or held until forfeited for nonpayment of taxes; but much of it for actual settlement. It is singular and significant of the characteristics of our institutions, to observe how small a proportion of the land now remains in the hands of the descendants of the original proprietors. A large proportion of it changed hands, during the first twenty years; and although the names sound familiar enough, it will be found on examination that but few of the present actual landholders of the Panhandle, are represented in the family names above recorded. In the mutations of circumstances, many who were then at the top of the wheel, have revolved downward; and while others, who were of more humble pretensions then, now occupy situations that enable them to look down upon others again, who at the next revolution may occupy their places. So it goes.

The easy character of the warrants, carelessness in locations, and the liability to be sold for taxes and purchased by speculators, caused a great deal of litigation in early times; and the land suits of that day were a perfect harvest to the attorneys, many of whom prospered and grew fat by nurturing and encouraging a litigious spirit among the settlers. There was no lack of the trading spirit among the settlers, as is evidenced by the frequent alienations, which seems to have amounted to a mania almost, about the year 1800, nor was there any deficiency of manufacturing enterprise. Previous to 1800, the manufacture of iron from the ore had been carried on successfully at the old furnace on Kings creek; and in 1801, James

Campbell conveyed the furnace with 300 acres of contiguous land, to Peter Tarr and James Rankin, for the consideration of $3,600 for the premises. The furnace was operated for many years afterwards, but has been now for a long time abandoned and in ruins.

After the year 1800, we enter upon the modern era.—The Indians, by this date, had been effectually expelled, towns and villages had sprung up at different eligible points; and population diffusing itself throughout the country, it rapidly lost its backwoods characteristics. Population increased with amazing rapidity west of the Ohio, and it was no unusual thing to see long trains of emigrant wagons, waiting their turn to be ferried over, at Wellsburg, Wheeling, and other crossing-places. Ohio became the Mecca of emigration, and the flood poured into her borders, enriching and fructifying the territory through which it rolled. Henceforward, the history of the country is that of a peaceful and thriving community, intent only upon the accumulation of wealth, the securing of worldly ease, and the fruition of the perils and hardships, encountered by our fathers. Though abounding in incident, it is not of that stirring character that will interest the reader.—The old pioneers became rapidly merged in the general mass of the population, and soon lost much of their distinctiveness of character. As illustrative of the physical capacity, of the men of that day, we give the following well authenticated incident, showing that they were big of body as well as of mind; and able to cope with the bears and Indians, as well as abundantly willing: In the year 1807, John Cox, then Sheriff of Brooke County empannelled a jury of twenty-four citizens, whose gross weight is recorded at 7230 pounds, or an average of 300 pounds each. It is probable that these were men of extraordinary size even for their day, or the empannelling of such a jury would not have been made the point of a newspaper paragraph, as it was; but there are few thinly settled countries, where half that number of as weighty characters can be found now, by the exercise of the greatest industry. Some of their names are given as follows; Mr. McGruder, Jas. Crawford,. Joseph Applegate, Francis M' Guire, Cornelius H. Gist, Jas. Connell, Amon Wells, Caleb Wells, Adam Wilson; James Robinson, Samuel Wilson. Lemon Fouts, Hezekiah Hyatt, and Absalom Wells, Sr. and Jr—three of them weighing near 400 lbs. each, and no man less than 240. The same account goes on to say, that at the same time could be counted on the waters of Short Creek twenty five or thirty ladies of corresponding dimensions, averaging from 240 to 300 lbs. avoirdupois. Such were some of the characteristics—social and physical of our pioneers; in subsequent chapters we will treat of their moral and intellectual history and of the material developments of the country. In neither respect is there much to regret or aught to feel ashamed of.

CHAPTER V.

RELIGIOUS CHARACTERISTICS.

Early Religions Inclinations—Intolerance—Presbyterianism—Sectarian Schools—Canonsburg College—Washington College, Pa.—Washington College, Va.—Seceders—Redstone Presbytery—Camp Meetings—Methodists—Persecution—Itineracy—Lorenzo Dow—Rev James Finley—John M'Dowell—Stone Meeting House on Short Creek—Rev. J. Monroe—Castlemau's Run Camp Ground—Baptist Denomination—Jonathau West—Alexander Campbell—Episcopal Church—Rev. Joseph Doddridge—Disputation.

There was early manifested a decided partiality for the forms and ordinances of christianity among the early settlers of the country of which we treat; in some portions of it, verging upon intolerance. The imigrants brought with them the peculiar religious tenets and inclination of the neighborhoods whence they came. That portion of the population which had its origin in Virginia and Maryland, was strongly tinctured with high church Episcopacy and Catholicism; the disciples of Wm. Penn were represented in the emigration from his province; while the strong Scotch Irish population, which so much proponderated in Western Pennsylvania, represented Presbyterianism, in every shape and form, as well as every phase almost of secession and reformation. Presbyterianism, positive or negative, in some shape or other, seems to have been the prevailing religion of Western Pennsylvania. Its missionaries were scattered all over the country, and were zealous in their labors; every opportunity was used by its colporteurs and ministers, to distribute bibles and tracts; they would visit emigrant boats descending the river, to see that their spiritual wants were duly attended to, and through the agency of missionary societies, take advantage of every opportunity to diffuse the gospel. The Rev. Mr. Patterson, alone, during fourteen years' residence in Pittsburgh, at this early day, in this way, distributed 6863 copies of bibles and testaments. They founded schools and colleges, and filled them with scholars, and supplied them with zealous and competent teachers.—In 1796, they resolved to establish two seminaries, in which the purpose of "educating young men for the gospel ministry," was a prominent object; one to be established in Rockbridge County, Va., under charge of Rev. Wm. Graham, as President, the other in Washington County, Pa., under care of Rev. John M' Millan. This was the origin of Washington College, Lexington, Va., and of Canonsburg College, in Washington County, Pa. Books of a doctrinal nature were enjoined to be put into the hands of the students, at once, on their entrance, indigent pious young men were provided for, and the two schools were placed under the supervision of a Board of Examiners, chosen from the Presbyteries respectively. A few years afterwards, Washington College, in Washington county, Pa., was instituted on similar principles.

The Presbyterian organization is essentially missionary. The printed records of the church, establish the fact that near one hundred years ago, she sent out missionaries into the howling wilderness west of the

Alleghenies to preach to the scattered emigrants, hunters, traders and indians who might fall in their way. As early as 1760, we read of their labors and travels in this capacity. Very many of the settlers of Washington and Allegheny counties, were seceders from the regular organization, and of the straitest sect of that persuasion. They were very dogmatical in their opinions and somewhat disposed to bigotry; much given to long sermons, very peculiar psalmody and cold meat on Sunday. Many of this denomination, may still be found in western Pennsylvania. The Presbyterian synod of New York and Philadelphia established in the year 1781, at the request of the Revs. Joseph Smith, John M'Millan, James Power and Thaddeus Dodd; what was called the Redstone Presbytery, which embraced the country lying between and upon the branches of the Monongahela and the Ohio; and took its name from Redstone Old Fort, which appears to have been a sort of head quarters, and gave the name of Redstone settlement to a wide extent of country. This Presbytery, was served by men of eminent piety and ability, among whom may be named--Revs. Joseph Smith, John M' Millan, James Power, Anderson, Dodds, Edgar and others—men who made their mark upon the early history of the country and the leaven of whose christian virtues, still works among the sturdy yeomanry of West Pennsylvania. The united congregations of Buffalo and Cross Creek united in a "call," it is said the first upon record west of the mountains, to the gentleman first named, in June 1779, promising the consideration per year, of £150 continental currency of 1774 for his services; and recapitulating the great loss "youth sustain by growing up without the stated means of grace; the formality likely to spread over the aged; and the great danger of ungodliness pervailing among both, there being divers denominations of people among us, who hold dangerous principles tending to mislead many weak and ignorant people; we cannot but renew our earnest entreaties that you will accept this, our hearty call." Houses of worship were extremely rare in those days, and it is said that none existed prior to 1790.—Even in the winter, meetings were held in the open air. A place was selected, as well sheltered from the weather as possible and a log pulpit erected; and in this primitive style the worship of God was conducted with as much decorum and perhaps with more acceptability, than in the gorgeous edifices and gilt edged pulpits of their descendants. This was the origin of the camp meetings, which were not, as is generally supposed, by any means confined to Methodists. They had their origin in the necessities of the country before Methodism existed; and were very generally adopted, not from choice, but for want of better accommodations.

Next in numbers and influence after Presbyterianism comes Methodism, though it does not by any means appear as efficient or at least as prominent, until of much later date. Indeed, in the early days of Methodism its professors and preachers appear to have been in very bad repute, and were considered rather as grievous nuisances to society, than as a reputable, christian denomination. Their more liberal and popularized doctrines and mode of church government came in direct conflict with the straight-laced Calvanism, so prevalent at that day; and as they commended themselves with more acceptability to the reckless, thoughtless and more ignorant masses of the community, Methodism became an object

GOING TO CHURCH IN OLD TIMES. — [Page 248.]

of jealousy, contempt and hatred. Methodism, under the preaching of Whitefield and Wesley in England had its rise and popularity chiefly among the humbler classes in that kingdom; its history was associated with many extravagancies, and with much that excited ridicule and reprehension; and the vulgar prejudice, excited against its preachers and professors, by the adherants of the English church, followed its ministrations across the Atlantic, and even into the wilds of the back woods. Nevertheless, there was at the bottom of its extravagancies, a solid stratum of truth, sincerity and pure piety that disarmed opposition; and the martyr-like devotion of its early preachers, recommended it to the masses, so that gradually it worked itself into notice, and became one of the leading denominations of the land. It appears emphatically, to have been the democratic church, in contradistinction to the more aristocratic and exclusive cotemporary sects. Commending itself to the sympathies of the masses and appealing rather to their feelings than to their intellects, it was the creed to prevail in a naturally consciencious, but uncultivated community, and the beatific vision, of supernal ecstacy into which its wrapt votaries were often inducted by overwrought imagination, or as they claimed, by the direct visitation of the Almighty, were of so impressive a character they could not only not be forgotten, but made them proof against all opprobrium and against all persecution. Itineracy was a peculiarity of the sect. The preachers emulated the example of the apostles in the simplicity and scantiness of their outfit. They took no thought of to-morrow, but depending upon the gospel staff and script, they relied upon what the day might bring forth, for their sustenance and support.—They dived into the bosom of the forests and tracked its almost pathless wilds; with a kind of spiritual knight errantry, they crossed unknown rivers, and plunged into dismal swamps--they came unawares upon the settler in his secluded cabin, and preaching with a zeal that would brook no denial, they used for his conversion sometimes carnal as well as spiritual weapons.—Where two or three could be gathered, they made the woods resound with prayer and praise and exhortation. There was a heroism, a self devotion, a defiance of peril, an endurance of hardship, and an obvious poverty, that vouched for their sincerity, and commended them to the respect and hospitality of their hosts. In this way, they sowed broadcast over the land, the seeds of Methodism, which were destined soon to grow up into a bountiful harvest. Among the first and most notable of these early itinerants was Lorenzo Dow, who gained a world-wide reputation for his eccentricities; and who first passed through this country about the year 1806, preaching at the different points on his route. He was not regularly in connection with the Methodist organization, but his doctrines had more similarity to theirs, than to those of any other denomination; and naturally he came to be regarded as a kind of apostle of Methodism. His travels commenced about the year 1792, and speaking of the sect in question, at that day, he says: "about this time there was much talk about the people called Methodists, who were lately come into the western part of New England. There were various reports and opinions concerning them, some saying they were the demons that were to come in the last days; that such a delusive spirit attended them that it was dangerous to have them speak, lest they should lead people out of the good old way they had been brought up in, that they would deceive if possible, the very elect." In his passage through this country in 1806, he speaks of preaching at Charlestown, and says that many were displeased with his preaching.—Returning again in 1813, he met with a kinder reception, at most of the points where he preached, though at West Middletown, Pa., he says that an effort was made to mob him, which failed. He was probably the first of the street preachers, and as often preached in the market place as in the church. He was possessed of much ready wit, which he could readily turn to advantage and very frequently to the ludicrous discomfiture of his antagonists and disturbers. Dow, was not the only Methodist preacher who was maltreated, nor was the prejudice against Methodists confined to particular localities. In Crawford county, Pa., in 1806, John McDowell, whose father's family was the first Methodist family in the county, preached the first sermon of the novel creed, and had almost to fly for his life from the vengeance of his congregation; as late as 1826, the Rev. Bear, who headed the first organization in Beaver county, was spit upon by the boys and otherwise insulted, during his sermon.

Rev. James Finley who flourished about the time of the last war, was an eminent preacher of this denomination, concerning whom, quite a number of anecdotes are afloat among his ancient friends. He appears to have been a kind of Boanerges—zealous, of powerful frame and utterly fearless, he would shake the souls of sinners over the fires of hell until they "squealed like young raccoons." He was a Kentuckian, but spent the greater part of his youth near Chilicothe, Ohio, and his father being a teacher of the classics, he acquired from him a superior education. He reproved sin without fear, favor or affection, and was not particularly careful of the phraseology he used. His rough practice brot' him frequently into disagreeable contact with the hard cases of his day. Said he, on being advised that a certain man in Steubenville, whom he had offended, had threatened to maltreat him. "I am willing to be led to the stake for the cause of Christ, but brethren, God never made the man who will ever cowhide James Finley." It is needless to say he was not cowhided, although he thundered his denunciations afterwards, with redoubled vim.

Nevertheless, and in spite of opprobrium and hostility, the church grew apace, and at an early day took rank with the Presbyterian in popularity; and in many, sections actually outstripped it in numbers. At this day it considerably exceeds any other denomination, in this section in the number of its members, and is behind none in popular estimation. One of the first organizations was established in the neighborhood of West Liberty, on Short Creek bottom, about the year 1805, and perhaps the oldest Methodist Church in the country is the old stone meeting house on Short Creek bottom, erected by them about the year 1810. Rev. Joshua Monroe, speaks of preaching in it in the year 1811, when it was in an unfinished condition, and states that the stone work was executed by Mr. Ralph Douglass, an Englishman and a Methodist of the old Wesleyan stamp, a sensible and deeply pious man who died a few years afterward in Washington, Pa. It is a venerable and time worn edifice, suggestive of old times; and surrounded with the grave stones of many of the patriarchs and pioneers of this section. Not far from it is the old Castleman's Run Camp Ground, also located about the same time (in 1814,) by the same Joshua Monroe, above mentioned, with others, laymen and preachers, and arranged for a camp ground. Prior to 1811, Camp Meeting had been held in the vicinity of the stone meeting house, but an intermission occurring at this time, the new site was selected on the land of the Jones family, and annual meetings have been held on the spot with great regularity, from that day to this.

Among the early Methodist preachers may be named Hoge, Page, West, John Waterman, J. Monroe, Jacob Young and others, many of whom will be remembered by some of our readers as men of great ability, piety and zeal in the cause of Methodistic Christianity. Those of them living now can look back upon the early days of their church and compare it with its present growth and strength with thankfulness to God, and honest pride at the commanding position it has attained to from such small beginnings.

The Baptist Church comes next in numerical importance in this section. It too, in infancy, had to encounter prejudices and sectarian hostility; but though divided into sects, it outgrew them all and attained a proportionate importance.

About the year 1801, Jonathan West of the county of Jefferson, N. W. Territory, deeded to the Regular Baptist Church of Kings Creek, Va., for the sum of one and a half dollars, sufficient land on which to erect a church. The church was afterwards erected, and for many years occupied, being among the very first edifices for such purposes in the western country. The Regular Baptists were afterwards divided into various sects, who discussed their various points of difference, with much zeal and ability. One of these sects or divisions, is that known as the Disciples or Campbellite, of which Alexander Campbell of Bethany College, may be considered the founder and exponent. A man of great industry, ability and zeal, he was in early life inbued with Calvanistic notions, but also with a free thinking and independent mind, and withal given to disputation. He early evinced a disposition to travel from the beaten paths, and originate a system peculiar to himself, which should embody his ideas of right christian doctrine and church government. His peculiar sect however, did not come much into vogue until a later day, and does not particularly come at this time within our view.

The following sketch of the life of Dr. Joseph Doddridge, whose "Notes on Virginia," have given his name a wide celebrity, will be found, also, an interesting sketch of the progress of the Episcopal church, in this region.

Prominent among the early citizens of Wellsburg, were the Rev. Dr. Joseph Doddridge and his brother, Philip Doddridge, Esq., both of whom attained to eminence in their professions. From early life, they were eager in the pursuit of knowledge, cheerfully expending their little patrimony in procuring, from abroad, that assistance which the paucity of instructors and books, at that early period, denied them at home.

Their father, John Doddridge, originally from Maryland, and a lineal descendant from the Rev. John Doddridge, of Shepperdton, England, emigrated at an early period of the settlement of the country, to the Western part of Washington County, Pennsylvania, locating in the neighborhood of the Virginia line. Being a man of piety and intelligence, although not enjoying robust health, he found many opportunities of rendering himself useful to the community in which he lived. When a resident of his native State, he was a member of the English Church, but after his removal to the West, he attached himself to the Wesleyan Methodists, for whose accommodation he erected, on his own farm, a house of worship, which still retains the cognomen of "Doddridge's Chapel."

Joseph, his oldest son, was born in October, 1768. At an early age, in Philadelphia, he took orders in the Protestant Episcopal Church, and during many years, labored, almost single-handed, in Western Virginia and Ohio, to collect and keep within the fold of that branch of the Church of Christ, its scattered members, not doubting that his brethren in the Atlantic States would early feel the importance of surmounting the great Allegheny barrier, and by their timely visits and affectionate christian ministrations, second and complete his efforts for the early and permanent establishment of the Episcopal Church in the western regions. But in this fondly cherished hope he was doomed to disappointment. Year after year passed, and still his oft repeated entreaties for help were only answered by plausible pretexts for delaying to a more opportune period the anxiously coveted assistance.

From the Hon. Judge Scott's reminiscence of the Rev. Dr. Doddridge, we learn that in 1793, he held regular Episcopal services in West Liberty, Virginia, then the seat of justice for Ohio County, and the residence of many respectable and influential families, most of whom removed to Wheeling, when the courts were transferred to that place. According to the same authority, St. John's parish, in Brooke County, seven miles from Wellsburg, was formed by him in the same year, and a small church erected. Of this parish, he continued the pastor until within a short period of his decease, when failing health compelled him to relinquish it.

In the year 1800, Dr. Doddridge formed a congregation in this place, then called Charlestown, also one in Jefferson County, Ohio, now known as St. James' church, on Cross Creek, in that County. As early as 1794 and '97 he held frequent religious services at Steubenville, Wheeling and Grave Creek.

In later years, his ministrations as a pioneer missionary were extended into the interior of the State of Ohio, and it was owing in a great measure to his zealous and persevering efforts that the preliminary steps were initiated which resulted in the erection of the state into an Episcopal diocese and the election of its first prelate, the energetic, self-denying and devoted Bishop Chase.

Some years after entering the ministry, the subject of this notice, in order to meet the wants of an increasing family found it necessary to combine with his clerical profession one that would be more lucrative in the region in which lie lived. He chose that of medicine, completing his course of preparation in the Medical Institute, of Philadelphia, under Dr. Benjamin Rush—In the latter profession he stood deservedly high, and to its avails he was mainly indebted for means to rear and educate a large family of children. But his practice being in a new and sparsely settled country, was laborious in the extreme, and laid the foundation for a disease which, in the latter years off his life, was painfully manifested.

In his disposition he was social and cheerful, fond of the society of friends, to whom he was always affable and accessible, aiming to his conversations with them to combine instruction with entertainment. His heart was replete with sympathy and compassion for the poor and the afflicted, to whose relief he ever imparted largely of his limited means. For some years previous to his decease lie was severely afflicted with an asthmatic complaint which finally terminated his life in the 58th. year of his age, in Nov. 1826. His remains with those of his parents, his wife and several of his children, repose in a monumental mound, in the old grave yard adjoining Brooke Academy, in this place.

In concluding this chapter on the religious peculiarities of the people of this section, we may safely say that no section of the Union can present a fairer record as relates to morality, and the elements of true religion; and few can be found where the leading tenets of christianity have been more thoroughly, zealously, and ably discussed. It has been the scene of zealous disputation almost from the time of its settlement, and if the disputants have now grounded arms, it is not from want of zeal, confidence or ability to dispute, but from the effects of a broader and wider spirit of christian tolerance, even to the verge of indifference. With a firm reliance upon the self-sustaining principle in christianity, we may hope that this kinder feeling may ever prevail until it merges in the consumation of the millenial hope, which all true christians are free to agree upon and indulge.

CHAPTER VI.

SCHOOLS AND COLLEGES.

Literary Tendency of the People—Quality Folks—Field Schools—Academies and High Schools—Alexander Campbell—Bethany College—West Liberty Academy—Wellsburg Female Seminary—Common Schools—Newspapers.

A People so eminently religious in their tendencies as were our forefathers, could not be indifferent to the education of their children; accordingly, we find great attention bestowed upon the education of youth. Considering the paucity of population, the inferior quality of the teachers, and the harassing nature of the times, it is as singular as it is ereditable, that education such as it was, was so general. The fact that it received so much consideration is in a great degree attributable to the character of the settlers themselves. At a very early period a class of settlers came in, who possessed a degree of refinement and intelligence, equal at least to any to he found in the sections whence they emigrated. Many of them contrived to gather around them the usual appendages of higher social life. Though their houses at first, were humble, often only a single log cabin, yet many of them owned slaves, possessed negro quarters as comfortable as their own, kept fine horses, and dispensing hospitality with a liberal profusion, essayed to be thought, what they were called by the less aspiring, "quality folks." These quality folks were generally well educated, and were both emulated and envied, by their less favored, but equally ambitious neighbors. Their own sons and daughters, they sent off to the east, to receive the polish of the college and seminary; while the others were encouraged to patronize the field school. The field school was an institution in its way. As described by writers of the day, and as some relics, now existing, prove, they were of the class of schools which benefit through much tribulation. A log house, of moderate size, was squatted down at the intersection of a couple of cow-path, or near some spring in the woods, the walls chinked with mud, and sticks, and stones; the roof and floor of clapboards, and doors, windows and chimnies, of the most primitive style. To this temple of learning, resorted the urchins for miles around,-- trudging through the woods in families; boys and girls together, with their dog-eared school-books, that had served the purposes of more than one generation. The teacher, some countryman of Ichabod Crane, or more probably, a gentleman from the bogs of "swate Ireland," who, by his blarney, induced his simple-minded patrons to believe him a paragon of "larnin" as well as a pattern of propriety, presided over this motley crew. The scholars sat bolt upright, on backless benches, while the *magister artis*, presiding with infinite majesty, kept them in terror of his rod and rule. What he taught them was the application of the birch; what they learned was what they could not help. In process of time, the scholar became inducted into the mysteries of the elements, graduated when he conquered the single rule of three, and took his first degree when he acquired "round hand writing." Jolly times, they were, at the old field schools—checkered like our lives with much of pleasure, much of pain. The riotous pleasure of boyhood, when released like young colts from durance vile, the warlike preparations of barring out, and the chivalrous punctilio of the assault, surrender and treaty, the juvenile gallantry of the youngsters toward the blushing lasses—all these recollections of old lang syne, as they rush back upon the memory, drown out the doleful hours of enforced quite, the painful confinement, the bothering of brain over intricate problems, the visions of the birch, smart of the bitch itself, and the manifold exacerbations of the youthful spirit. Reminiscences such as these, and many more, balance each other on memory's chart, as the mind recurs to the school boy days.

The teachers themselves, were not generally very advanced in learning, nor were they always given to habits of strict sobriety; were poor in purse, and often compelled to eke out a scanty livelihood by other avocation; among which were those of travelling cobbler and tailor. They boarded round among their patrons, and in the intervals between their professional engagements, they mended tire shoes and made the breeches for the families with whom they boarded.

These field schools, as they were called, existed until a late day, and indeed are not yet extinct; but as the population increased, the standard of education was advanced and academies and colleges were planted at various points, wherever numbers justified.

Toward the close of the last century, a movement was made by the members of the Redstone

Presbytery, to establish schools on a firm foundation, having in view the advancement of the church. In 1792, an academy was established at Canonsburg, in Washington co. Pa., and another in Lexington, Rockbridge county, Va., believed to be the first high schools west of the Allegheny mountains. These academies or seminaries as they were called, were kept up for a time by contributions from the people, but in a short time, they received sufficient patronage to be self-supporting. The Seminary at Canonsburg, was converted into Canonsburg College in 1802, since which time, the institution has grown in popular estimation and become one of the most respectable institutions of learning in the Union. About the same time that Canonsburg College was established, a competitor sprung up in Washington, which after a chrysalis existence of a few years, assumed the name of Washington College, and in time attained to great prosperity. Academies and high schools multiplied and kept pace with the progress of the country. At every considerable town, an Academy was established and sustained—, sometimes aided by private munificence, sometimes assisted by the State, and at others depending solely upon their merits for their support. An academy was established at Wellsburg at an early day, which furnished means of education to many now in active life, and once possessed considerable celebrity. Alexander Campbell also, was early distinguished as an instructor of youth, many of the middle aged citizens of the present day having received their education under him. His career as an instructor, culminated in the establishment of Bethany College in 1840.

Mr. Campbell, was born at Shane's Castle, Ireland, in 1778, and having received a finished education is his native country, he emigrated to America in 1808. He located shortly after his arrival, in Western Pennsylvania, in the capacity of preacher and teacher; and noon acquired a reputation as a man of talent and ambition. He was a delegate to the convention to reform the Constitution of Virginia in 1829-30, but never particularly distinguished himself in politics; his principal forte being in controversial writing and debate.—The Millenial Harbinger, established during the year 1823, of which he has continued the principal editor and always the leading spirit, has exerted a great influence on the religious sentiment of christendom; and he has engaged in various public discussions in this country and in England, which have given him a reputation almost cosmopolitan. He has also edited and published several books of a theological character. Few men living, have, indeed, transacted so much or such diversified business as Mr. Campbell; or labored through life with such unremitting industry. He has accumulated considerable property and his homestead at Bethany, Brooke County, Va., is the scene of profuse hospitality to visitors from all sections of the country attracted by the wide spread reputation of the man, as well as by the calls of business, religious, literary and secular. In personal appearance, he is tall, venerable and dignified; and the most casual observer, would not fail to recognise in him, the marks of a commanding intellect.

  He early showed a disposition to differ from the Calvanistic preachers among whom he was thrown, and after much disputation, controversy, and even abuse, he left them and essayed to establish a creed and practice more in accordance with his own ideas of scriptural propriety. In this, he has to a great extent succeeded and he is now, at least by the public, considered the head of the branch of the Baptist denomination, which has taken in some sections, his name. For himself, in all his teachings and writings, he emphatically disclaims sectarianism; but popular opinion, seems to judge him differently from his own judgement of himself.

For the last twenty years of his life, the engrossing object of his attention, has been to establish a college near his homestead at Bethany, where his ideas of christian culture may be appropriately developed.

The first definite plan of the organization of the College is laid down by Mr. Campbell in the Millenial Harbinger for October, 1839, under the head "New Institution." The project had been long ripening in his mind, but its realization had been deferred waiting the successful establishment of Bacon College, Kentucky, in the success of which, Mr. Campbell, took a great interest. His first idea was, that the location of the college should be "entirely rural—in the country, detached from all external society; not convenient to any town or place of rendezvous—in the midst of forests, fields and gardens—salubrious air, pure water—diversified scenery of hills and vallies, limpid brooks and meandering streams of rapid flowing water. Such is the spot I have selected." This description sounds somewhat Acadian, but it correctly delineates the landscape, while the event shows that Academic seclusion has proven a failure—a thriving village springing up around the very doors of his college.

His next grand idea was, that the school should be free from any sectarian influence, and thus severed from the dogmas of established religions, induct more rational theology into the minds of students than he deemed to then prevail. Says he: "We want no scholastic or traditionary theology. We desire, however, a much more intimate, critical, and thorough knowledge of the Bible, the whole Bible as the Book of God--the Book of Life and of human destiny, than is usually or indeed can be, obtained in what are called Theological Schools."

His model school was to he built up on an original plan combining in its detail the requisites both of church and college and of church, preminently. To quote farther from his programme. "This church institution shall, in one cardinal point of view, resemble the West Point military school. There, it is not the theory alone, but the military camp, the practice, the daily discipline of the god of war. In this institution it will not be the theory of a church—of Bible reading, Bible-criticism, Bible-lectures—sermons—church order—Christian discipline; but daily practice of these. This church will be session seven days in every week.—The superintendant of this institution, or the professor in attendance, will be bishop *pro tempore* of the church. The young men, in all their readings, questions and answers, and exercises, shall rise, and speak, and act, its though they were, as in truth they are, members of a particular church met for edification and worship."

His programme then goes into detail of prospective arrangements, some of which have been consummated and others proved visionary. The College was incorporated in 1840. The second annual meeting of the Trustees was held at Bethany, on Monday May 10th, 1841, at which time, the available funds of the Institution were stated at $11,054, obtained by subscription, a considerable portion of which was by Mr. Campbell himself, who was acting as treasurer and agent. Four Professors had been nominated, two of whom, W. K. Pendleton, a graduate of the University of Va., and Robert R. Richardson, M.D., accepted their appointments as stated at this meeting. The bill of fare at the Stewart's Inn, it was resolved, should be the same as at the University, and the cost of one year's attendance, was unanimously fixed at $150; besides an entrance fee of $10, for each student.

At this time, the buildings were unfinished, and but a little over $1000 of the subscriptions actually paid, although the Inn and the College were being built.—The Treasurer asked $20,000 from the community and in consideration, promised not only an abundance of competent instructors, but accommodations for five hundred students. To raise the requisite funds, he labored with an assiduity the most indefatigable, and travelled into the most remote sections of the Union, and even beyond. For the first few years of its existence, the College struggled against manifold difficulties, not the least of which was opposition on sectarian grounds, but finally, it overcome them all, and, at this day, realizes, in some degree, the anticipations of its venerable founder and President.

Notwithstanding his religious peculiarities, the reputation of Mr. Campbell attracted an indiscriminate patronage, and gradually his school worked itself, not only into notice, but into the possession of considerable patronage. The Chairs of several Professors are now endowed, in sums sufficient to afford them comfortable salaries, and are generally filled, and with men of the first order. The Old College building, which was accidentally burned, in December, 1857, was replaced the succeeding season, by a portion of the present magnificent edifice, dedicated December 10th, 1858, the funds having been obtained by subscription, among those friendly to the Institution. The building destroyed was not of much value; but the valuable libraries, cabinets, &c., belonging to the College, some of which it will be impossible to replace, were a serious loss.

West Liberty Academy, established under the auspices of Prof. A. F. Ross, until the spring of 1858, a professor in Bethany College; and under an old act of incorporation, assisted by the State, commenced its first session, August, 1858.

The Female Seminary at Wellsburg, was established in 1852, professedly to be under the patronage of the Methodist E. Church, but although agents were put in the field, a considerable sum raised, and one wing of the edifice erected and occupied, it has not yet been completed. Colleges, Female Seminaries, and Theological Institutions abound throughout this section.

Nor, in the anxiety to build colleges and establish seminaries, have the people been unmindful of humbler educational wants. The State of Pennsylvania early established a Free School system on a magnificent basis, and in Washington county, their common schools have ever been an object of pride as well as of attention. Common School teaching has been reduced to a science and systematized almost to perfection. A magnificent edifice for the purpose of a Union Free School was erected in the town of Washington, in the years 1855-6, at a cost of some $20,000, and comfortable school-houses are thickly scattered throughout the borders of the county. In Virginia, the law allows counties that see fit to do so, to adopt a Free School system, similar in its provisions to that of Pennsylvania. The county of Ohio adopted it about the year 1852, by election. A considerable amount of money was invested in schoolhouses, but the system does not seem to operate so satisfactorily as could be desired; in Brooke county, at the same election, Free Schools were voted down by a small majority; and in Hancock and Marshall, subsequent elections have resulted similarly. Under the general law of Virginia, which makes quite liberal provision for Common School education, though clogged with provisions which render it distasteful to the classes it is intended to benefit, the facilities for acquiring a common school education are good, and where there is a disposition, there is abundant opportunity. The proportion of persons unable to read and write, is smaller in the Panhandle, than in any other section of the State, even with the present unpopular and radically defective system.

Take all things into consideration, and no section can be found in the Union, surpassing this in the morality, intelligence, law abiding spirit and general competence of its inhabitants, a fact not more honorable in the present generation than creditable to their progenitors.

CHAPTER VII.

INTERNAL IMPROVEMENTS.

The construction of a wagon road from Will's Creek to the Ohio was early an object of solicitude on the part of the Government and people of the country. As far back as 1748, Thos. Walker, Thos. Rutherford, Jas. Wood and Abram Kite, Gent, or any two of them, were authorized and empowered by the Colonial Assembly to lay out a road from the North branch of the Potomac to Fort Pitt and for the furtherance of the object, the sum of £200 was appropriated.

The reason assigned for this enterprise was that an advantageous trade might thus be opened with the Indians on the western borders "of this dominion," and the King's garrison be the better supplied with provisions. They were instructed to follow as near as might be, the route of Gen. Braddock in his ill fated expedition of 1755, and the result of their explorations was the road for a long time used and finally adopted with a few variations, as far as the Monongahela, as the route of the National Road. As the population increased, it demanded an improved connection with the East. Toward the close of the last century, emigration poured over the mountains in almost a continuous stream; and in pursuance of the policy of the government to foster the settlement of the great west, the scheme of a great National Road from Philadelphia to the Ohio, and thence traversing the Northwest Territory to St. Louis, or the mouth of the Missouri, was projected, and soon became the subject of much discussion both in and out of Congress. When the State of Ohio applied for admission into the Union in 1802, she was admitted with the proviso, that one twentieth part of the public lands within her boundaries should be set apart that the proceeds might go to the construction of such a road through Ohio and ultimately to St. Louis.

On the 29th March 1806, Congress passed a law providing for the construction of the road from Cumberland to the Ohio, and Thomas Moors of Maryland, Joseph Kerr and Eli Wilson of Ohio, were appointed Commissioners to decide upon a route. The route proposed by them with only one deviation at Uniontown, was approved by President Jefferson in 1808, as far as Brownsville—the route, from that point to the Ohio, being left undetermined. The point at which the road would strike the Ohio, was considered as of the utmost local importance, and every eligible point on the Ohio, from Pittsburgh to below Wheeling, was warmly engaged in urging its claims. It was anticipated that a city would at once spring up wherever the crossing was definitely fixed. At this period, dates the jealousy that subsequently existed between Wheeling and Pittsburgh; and in a greater or less degree with all the other points on the eastern shore of the river. It became a delicate question for the commissioners to decide, and remarking that, "in this, was to be consulted the wishes of that populous section of Ohio, and the connections with roads leading to St. Louis, under act of 1806," they left the question open. The route from Brownsville, to Wheeling, was afterwards located by another commission, the engineer for whom, was a Mr. Weaver. Operations on the road were commenced forthwith and up to 1817, it had cost $1,800,000, and had moreover in some portions become worn out so as to need extensive repairs. The question of abandonment came up. In 1822, President Monroe issued his celebrated Internal Improvement message, in which he argues with consumate ability the general improvement policy of the country, and enlarges upon the propriety of the government carrying out the original compact with the State of Ohio, by continuing the road west of the river Ohio. Three Commissioners, had been appointed in 1817, to locate the western division; and it is at this date than we first read of its Ohio terminus being definitely fixed at Wheeling. Col. Moses Shepherd, was a principal contractor on the road between Wheeling and Cumberland, Messrs. John McClure, Dan'l. Steenrod and others, had contracts more contiguous to the former place. The work was executed promptly and with apparent faithfulness; but subsequently, much litigation arose on account of alleged failure to comply with the terms of contracts in executing masonry, &c., which afterwards found its way into Congress in the shape of Bills for the relief of different parties. A large amount of money was expended by the government, and large fortunes were made by some of the contractors out of the proceeds. The road gained great celebrity at the time from its magnificence of design, costly character, the romantic country traversed and the immense trade and travel that passed constantly over it. It became the grand artery of emigration as well as of transportation between the East and the West. Forty wagons in a train all magnificently belled and otherwise equipped, might be seen at one time traversing this national highway, loaded with merchandise for the whole country, as far west as St. Louis.

Notwithstanding, however, the immense travel and trade, the tolls were insufficient to keep it in proper repair, and bidding fair to become a burden on the Federal Treasury, a growing disposition was manifested to abandon or rather to transfer it to the States it traversed. About the year 1825, it was terribly out of repair, especially that portion of it between Brownsville and Wheeling; and so desperate had become the condition of the Western division that a change of location was seriously talked of from the Wheeling route to the route via Wellsburg. During the previous long and acrimonious contest for the crossing place, Wellsburg had been the equal and formidable rival of Wheeling, and now, when it was re-opened, she renewed her rivalry with a desperate zeal. Topographical advantages were confessedly in her favor both as to distance and nature of the ground to be traversed in order to strike the Ohio; but even at that early day and indeed a long time previously, the narrowness of the river had suggested the practicability of a bridge at Wheeling Island, and there were influences also on the Ohio side, that operated strongly in her favor. She was also fortunate in her advocates in Congress. Henry Clay, the reputed father of the internal improvement policy of the government threw in her favor the weight of his influence; and contributed greatly to her success by his zeal and his sarcastic allusions to "Panther Mountain," a high hill two miles to the east of Wellsburg which he came out of his way to explore on one of his journeys to Washington City, purposely to see for himself the comparative merits of the rival routes. He, perhaps, unwittingly, misrepresented the character of the Wellsburg route, the entire 23 miles of which, it has been estimated since, would have cost less than the two miles nearest Wheeling, of the route as adopted. But superior management triumphed and the original location to Wheeling was confirmed. When afterwards, Henry Clay became a candidate for the Presidency in opposition to Gen. Jackson in 1832, he was remembered by the adherents of the respective routes. Ohio county went for him with the greatest unanimity; while in Brooke, he only received one vote, that of Prov. Mounts, an eccentric, hair brained individual, whose solitary vote was for a long time a subject of amusement among his neighbors and acquaintances. The fact coming to the ears of Mr. Clay, elicited from him a humorous and good natured remark. Harry of the West was defeated; but the impress of this local controversy remained not only upon the neighborly relations of the parties but upon their political complexion. Wheeling, became thoroughly and persistently Whig; and together with the county of Ohio, firmly devoted to the interests of Mr. Clay; while Wellsburg, and all the vicinity sympathising with her, became uncompromisingly anti-Clay and Democratic. Subsequent events and the obliterating effects of time have softened and modified this local antipathy in some degree; but to this day, the effects may still be distinctly seen, both in local jealousy and national politics. At the time of this last desperate effort to wrest from Wheeling the possession of the terminus of the National Road, Phillip Doddridge, represented this District in Congress. This was in 1829-32. He was a resident of Wellsburg, where his talents were looked upon with the highest admiration, and where the highest anticipations were entertained of his acknowledged abilities and influence being exercised in favor of his native place. But the controversy seems to have so degenerated that no reasonable expectation could be entertained of a change of the location of the road; and however much Mr. Doddridge may have been disposed to favor his townsmen and immediate constituents, the margin for a plausible case and for a successful effort, was so extremely small, that he, perhaps, never seriously, entertained the hope of success, or, perhaps, the design of attempting it. Mr. Doddridge was a man of great liberality of views, there was very little of the contracted politician in his character, he took wide and national views of all subjects, and disdained to allow local considerations, however plausible, to influence his actions as a legislator. Such was his general character, and this, added to his rare colloquial powers, and great simplicity of manners, was the secret of his wide and universal popularity. In this case, he doubtless felt constrained to flatter his townsmen with some promise of success, but it is doubtful whether he ever entered fully into their designs. At any rate, he allowed the matter to go by default. Gen. Connell came on to Washington City, with recommendatory documents, signed by the citizens, but both the General and Mr. Doddridge got on a frolic together; and it is said, that the documents were never presented. The affair subjected Mr. Doddridge to considerable animadversion. From this, a knowledge of his character, is his best defence. He was long and extensively known, and admired as a jurist and statesman, his discriminating and comprehensive judgement in fathoming abstruse and intricate cases, as well as his powerful and logical arguments in elucidating them, having gained him unbounded popularity as an advocate. To his other advantages, were added colloquial powers of the highest order, which, being combined with extreme simplicity of manner; rendered him, at all times, a most fascinating and interesting companion. He was born near Philadelphia, in May, 1773; came to the West about the year 1785, with his father's family; and, at an early age, applied himself assiduously to the study of the law. He, and his brother Joseph, were in a great degree self taught, and rose to distinction by force of industry and native vigor of mind. For several successive session he represented Brooke County, in the Virginia Legislature; and there, distinguished himself as well by the soundness of his views as by his commanding eloquence.

In 1828, he was elected to Congress after a hotly contested canvass, took his seat, March 4th 1829, at the beginning of Gen. Jackson's first administration, to which he was opposed, voting with the majority, for the recharter of the United States Bank—the absorbing issue of the day. Near the end of his term, June, 1832, he died suddenly, and lies interred in the congressional burying ground at Washington City. He left a widow and ten children.

He was but a poor financier, and left little else to his heirs except his memory, which is idolized by his family and embalmed in the hearts of his many warns friends and admirers.

In this connection it may be well enough to give some history of the Wellsburg and Washington Turnpike, which was originally intended, if not to take the place of that portion of the National road extending from Washington to Wheeling, at least to divert at the former place some portion of the stream of travel in the direction of Wellsburg. It is a contemporary of the Cumberland road, and is one of the very oldest macadamised roads west of the Allegheny mountains. The original charter was passed in 1808. It commences in rather grandiloquent style by reciting that it "is contemplated to build a continuous highway from the city of Philadelphia and from the 'Potomac' river, to Charlestown, to intersect the Federal Highway from the Potomac to the Ohio, at some point, between Washington and Brownsville, Pa. Books of subscription were authorised to be opened and Col. James Marshal, Oliver Brown, Moses Congleton, John Connel, N. P. Tilinghast and James Perry were named commissioner. The capital stock was to be $15,000, divided into shares of $50 each and it was specially provided, that all excess of profit over 15 per cent, was to be applied as a sinking fund for paying back the stock of the road. Nothing, however, appears to have been done under the charter until about the time when the National Road had come into such bad repute for the want of repair, that there was a prospect of its abandonment from Washington to Wheeling. The possibility of the Wellsburg route being adopted in that case, encouraged the corporators again to open their books. Col. James Marshall a man of great enterprise and public spirit appears to have been particularly active. About the year 1825, stock was subscribed and the route surveyed and the road actually put under contract. Considerable work was done on it, but public opinion was too strongly in favor of the Wheeling route; the Pennsylvanians failed to second the efforts of their Virginia neighbors and on the event of the road being finally confirmed to Wheeling as stated above, the project was almost abandoned in despair. The road languished for some years afterwards, but was gradually put into good condition and although the original design was a failure, and the 15 per cent profit was never realized, still it has been of incalculable benefit in opening up the section of country it traverses and affording a convenient outlet to the river for the western half of Washington county.

The National Road was finally relinquished to the States in 1836, after having cost the country some $7,500,000, in its construction and support. Just previous to this final relinquishment, the sum of $300,000 was appropriated to put it in good repair east of the Ohio, with the understanding that after its relinquishment, the general government was to be released from all further obligation on its account since that time it has been gradually falling into disuse. Railroads have changed the courses of trade, and the manner of travel. The rumbling lines of coaches, that used to career along its dusty stretches, have disappeared, with their army of Jehus; the ponderous roadsters have "gone to rack," grass grows on the road bed, and the villages and tavern stands that lined the road and lived off its droppings have fallen into decay. The National Road has had its day, and now does menial services as a country road for neighborhood accommodation. It was a magnificent conception at the time, and answered a magnificent end. It contributed more than any other one thing, to the rapid settlement of the west. It paid back with interest every dollar ever expended upon it.

THE WHISKEY INSURRECTION.

Toward the latter part of the last Century, occurred the Whiskey Insurrection—an event, for the history of which, we have but little space, but which occupied at the time a very considerable place in the public mind. In the early days of the Union, it was deemed expedient to lay a tax, per gallon, on all home-made spirituous liquors to help meet the pressing exigencies of the country. The general murmur thus occasioned, gradually subsided, except in the western part of Pennsylvania; and the region generally, of which we have been treating. The Scotch Irish element, was lashed into rebellion by the attempt to interfere with their cherished beverage and at the same time their main article of trade.

The country at that time had no reliable market nearer than New Orleans; and whiskey was the most economical commodity by the sale of which the settlers could realize money for their surplus grain. It was always saleable, not very bulky, and brought the ready cash.—Almost every spring and clear running stream had a still by its side, and scarcely a farmer, but was also a distiller. Grain, for the ordinary purposes for which it is used, was a drug; hard money was very scarce—12 1/2 cts. being often the extreme price for a bushel of wheat. The tax under these circumstances operated upon them with peculiar hardship, and, accordingly, when the collectors came round, their demands were refused. Matters progressed, until they ended in open mutiny.—July 14th, 1794, the insurgents, to the number of several hundred, surrounded the dwelling of John Neville, Inspector of Revenues for Fourth Pennsylvania district, seized upon his papers, destroyed his private property, and maltreated and wounded him and his servants. The consequence was, a complaint to the county authorities, but they being unable to protect him, he fled the vicinity. David Lenox, the U. S. Marshal, was similarly served. A proclamation soon appeared from the President of the U. S., cautioning the malcontents against the consequences of their treasonable acts, ordering them to disperse previous to the 1st of the ensuing September, and providing for the calling out of the militia for the purpose of enforcing obedience. This was at the instance of Jas. Wilson, Esq., associate Justice, who notified the President on the 4th of August, that combinations then existed too powerful for the ordinary process of law, and called for military assistance.

This proclamation was disregarded by the Insurgents, and on the 25th Sept., it was followed by another, advising them that troops were embodied and on their march to the disaffected region; but still offering amnesty to all disposed to claim it. The language of the President was strong, indignant, yet dignified, and backed by the overwhelming force that rallied to the support of the law, struck terror into the hearts of the leaders. They fled the country or lurked about in disguise, while their organization rapidly melted away, so that on the approach of the troops, who numbered 15,000 men, accompanied by Washington himself, as far as Carlisle, the Whiskey Boys, had become utterly invisible. The troops, committed many petty depredation, upon the property of the inhabitants, which were promptly indemnified by the Government; and the Whiskey Rebellion was ended without bloodshed, by the wise exhibition of such an overwhelming force as was sent out to suppress it.

Some of the ringleaders were arrested and imprisoned but the charges were never pressed; and a general pardon was extended to all, with a few exceptions, on the sole condition, that they would thereafter obey the laws as good citizens should. This leniency was wise and well timed. It restored many otherwise good citizens to their homes and the practice of industry;--while it convinced the disaffected, that the government while abundantly able to coerce obedience, was yet willing and disposed to deal fairly and kindly.

A small body of troops remained on the ground for a short period but no further disturbance occurring, they were removed, and the Whiskey Insurrection ended.

ADAM POE AND BIG-FOOT.

The mouth of Tomlinson's Run, in Hancock county, Va., was the battle ground of the celebrated Adam Poe and Big-foot Indian fight, the precise location of which, has never been exactly stated by the historians of that encounter. We give this, on the authority of Mr. John Brown, an old citizen, whose dwelling occupies nearly the identical spot, corroborated by the evidence of many others, who were cognizant of the fact from personal knowledge. Mr. Brown, possesses many Indian relies and takes pleasure in pointing out the spot and narrating his recollections of Indian times.—Some years ago, he found under some rocks a bark canoe, in a good state of preservation, which it requires no stretch of the imagination to presume, was the identical one in which the Big-foot brothers crossed the Ohio, on their last marauding expedition. The tale of the Poe fight has been so often and so well told, that we will not repeat it here, but our correspondent gives some additions which way prove interesting. The information is derived from Mr. Thomas Edgington, for two years a captive among the Indians. He was captured, when on his way from his cabin at the mouth of Harmon's Creek, to Col. Jas. Brown's Fort, to borrow of him a log chain. The Indians came suddenly upon him, made signs to him to surrender, but essaying to escape by running, he was mired in the creek, and they took him prisoner, hurrying him with them over the river and on to the Indian Towns. Simon Girty happened at the towns afterwards and through him, he ascertained that the Indian, whose prize he was, no other than the surviving brother of the Big-foot fight—bearing on his hand the scar of a severe wound, there received. The Indian stated, that on finding himself disabled by this wound, he stole away from the fight and swimming the river hid in the bushes until dark. He then constructed a raft recrossed the river, and recovering the bodies of his slain brothers, except that of the one who floated off, as narrated by the whites, he conveyed them to the Ohio side and there interred them. He then, being wounded and the last of five stout brothers, took up his sorrowful way back to his tribe, where their deaths were sorely lamented for many days.

Mr. Edgington paid a high tribute to Indian virtue in his description of this warrior. According to his account he was the "noblest, best man—the man of the host principle, he ever knew—white, black or red." Sometimes the other Indians would impose upon the captive. His master would pat him on the back to encourage him to fight, and would applaud his manly resistance. Sometimes when they would double on him, his captor would interfere with knife and hatchet, arid cut and slash right and left. He would share with him his blanket and robe, giving Edgington, the largest share, and divide with him his last morsel of meat.

Edgington, was finally released and returned home after a two year's captivity, but always held in grateful remembrance his kind hearted Indian master.

Mr. Brown, communicates another incident in Indian history, for which, we regret we have not space.